Materials Science and Engineering

Materials Science and Engineering

Editor

Shiw Gupta

Materials Science and Engineering

Edited by **Shiw Gupta**

Printed in 2017

ISBN: 978-1-68117-223-1

Library of Congress Control Number: 2015936672

© 2016 by
SCITUS Academics LLC,
616, Corporate Way, Suite 2, 4766,
Valley Cottage, NY 10989

www.scitusacademics.com

Contents

Preface...vii

Chapter 1   Microstructure Evolution and Mechanical Properties
              Improvement in Liquid-Phase-Sintered Hydroxyapatite by Laser
              Sintering...1
              Songlin Duan, Pei Feng, Chengde Gao, Tao Xiao, Kun Yu,
              Cijun Shuai, and Shuping Peng

Chapter 2   BIM Based Collaborative and Interactive Design Process Using
              Computer Game Engine for General End-users 21
              Gareth Edwards, Haijiang Li, and Bin Wang

Chapter 3   Approaches for Modelling the Energy Flow in Food Chains 55
              Baboo Lesh Gowreesunker and Savvas A Tassou

Chapter 4   Intrusion Detection and Big Heterogeneous Data: A Survey 97
              Richard Zuech, Taghi M Khoshgoftaar, and Randall Wald

Chapter 5   Mode I Critical Fracture Energy of Adhesively Bonded Joints
              between Glass Fibers Reinforced Thermoplastics 165
              Siripong Mahaphasukwat, Kazumasa Shimamoto, Shota Hayashida,
              Yu Sekiguchi, and Chiaki Sato

Chapter 6   Direct Patterning of Gold Nanoparticles Using Flexographic
              Printing for Biosensing Applications...187
              Jamie Benson, Chung Man Fung, Jonathan Stephen Lloyd, Davide
              Deganello, Nathan Andrew Smith, and Kar Seng Teng

Chapter 7   Biosensor for Human IgE Detection Using Shear-mode FBAR
              Devices 207
              Ying-Chung Chen, Wei-Che Shih, Wei-Tsai Chang, Chun-Hung
              Yang, Kuo-Sheng Kao, and Chien-Chuan Cheng

Chapter 8   **Angular Shaping of Fluorescence from Synthetic Opal-based Photonic Crystal**..227

Vitalii Boiko, Galyna Dovbeshko, Leonid Dolgov, Valter Kiisk, Ilmo Sildos, Ardi Loot, and Vladimir Gorelik

Citations...243
Index...247

Preface

Materials science or materials engineering is an interdisciplinary field involving the properties of matter and its applications to various areas of science and engineering. This science investigates the relationship between the structure of materials at atomic or molecular scales and their macroscopic properties. It includes elements of applied physics and chemistry. With significant media attention focused on nanoscience and nanotechnology in recent years, materials science has been propelled to the forefront at many universities. It is also an important part of forensic engineering and failure analysis. The material science also deals with fundamental properties and characteristics of material.

Editor

Chapter 1

Microstructure Evolution and Mechanical Properties Improvement in Liquid-Phase-Sintered Hydroxyapatite by Laser Sintering

Songlin Duan[1, 2], Pei Feng[2], Chengde Gao[2], Tao Xiao[3, 4], Kun Yu[5], Cijun Shuai[2, 3], and Shuping Peng[1, 6]

[1]Hunan Provincial Tumor Hospital and the Affiliated Tumor Hospital of Xiangya School of Medicine, Central South University, Changsha 410013, China

[2]State Key Laboratory of High Performance Complex Manufacturing, Central South University, Changsha 410083, China

[3]Orthopedic Biomedical Materials Institute, Central South University, Changsha 410083, China

[4]Department of Orthopedics, the Second Xiangya Hospital, Central South University, Changsha 410011, China

[5]School of Materials Science and Engineering, Central South University, Changsha 410083, China

[6]School of Basic Medical Science, Central South University, Changsha 410078, China

ABSTRACT

$CaO-Al_2O_3-SiO_2$ (CAS) as a liquid phase was introduced into hydroxyapatite (HAp) to prepare bone scaffolds. The effects of the CAS content (1, 2, 3, 4 and 5 wt%) on microstructure and mechanical properties of HAp ceramics were investigated. The optimal compression strength, fracture toughness and Vickers hardness reached 22.22 MPa, 1.68 $MPa \cdot m^{1/2}$ and 4.47 GPa when 3 wt% CAS was added, which were increased by 105%, 63% and 11% compared with those of HAp ceramics without CAS, respectively. The improvement of the mechanical properties was attributed to the improved densification, which was caused by the solid particle to rearrange during liquid phase sintering. Moreover, simulated body fluid (SBF) study indicated the HAp ceramics could maintain the mechanical properties and form a bone-like apatite layer when they were immersed in SBF. Cell culture was used to evaluate biocompatibility of the HAp ceramics. The results demonstrated MG-63 cells adhered and spread well.

INTRODUCTION

Bone grafts are frequently required to repair skeletal defects due to disease, trauma and congenital defects [1, 2]. Bone scaffolds are one of the three key elements for bone grafts [3]. There is an urgent need in bone repair field to develop biological applicable scaffolds, which could guide bone regeneration in defects [4]. To fulfill bone repair requirements, scaffolds should possess bioactivity, biocompatibility and porous structure, and provide a structural support for the growth of cells and regeneration of tissues [5,6,7,8,9].

Hydroxyapatite [$Ca_{10}(PO_4)_6(OH)_2$, HAp] with solid [10,11] or hollow shape [12,13,14] is widely used calcium phosphate as a scaffold material in clinical application because of its similar mineral component to the human nature bone [15,16,17,18]. Several recent studies have been performed and show that HAp ceramics possesses excellent bioactivity and biocompatibility [10,11,12,19]. Besides, selective laser sintering (SLS) is an effective way to develop porous scaffolds with an intricate and controllable internal structure [4,20]. Moreover, it is hard to achieve a high densification by using SLS because of the short interaction time (0.2–200 ms) between the laser beam and material powders in the process of sintering. In general, the densification has direct effects on the mechanical properties of ceramics [21].

Liquid phase sintering (LPS) is an effective approach to improve the densification [22]. It involves three overlapping processes: (a) crystallite rearrangement; (b) coarsening by solution-reprecipitation; and (c) grain coalescence [23]. In the processes, a close spatial arrangement is obtained due to the formation of capillary force [24]. In recent years, LPS has been used to improve the mechanical properties of ceramics. Mikinori Hotta $et\ al.$ [25] sintered SiC ceramics using $AlNY_2O_3$ as a liquid phase and found that flexural strength was increased to 1000 MPa by improving the densification. Defu Liu $et\ al.$ [26] reported that the fracture toughness and compression strength of β-TCP scaffolds were increased by 18.18% and 4.45% using poly-L-lactic acid (PLLA) as a liquid phase, respectively, compared with those of β-TCP scaffolds without PLLA. Uma Batra $et\ al.$ [27] reported that the sintered density and compression strength of HAp composites were improved using $CaO-Na_2O-P_2O_5$-based additives as liquid phase.

Various oxides such as calcium oxide (CaO), alumina (Al_2O_3) or silicon oxide (SiO_2) are the representative additives that have been widely used in HAp based ceramics [28,29]. The presence of small amounts of CaO, Al_2O_3 and SiO_2, which form liquid phases in the grain boundaries, can affect sintering behavior and thus enhance the densification [30]. Moreover, silicon (Si) ion is beneficial to cells growth, mineralization of calcined tissues and bone calcification [31]. In this paper, the $CaO-Al_2O_3-SiO_2$ (CAS) ternary-oxide system was selected as a liquid phase to sinter the HAp ceramics. The effects of CAS content on the microstructure and mechanical properties of HAp ceramics were investigated. The $in\ vitro$ bioactivity and biocompatibility were evaluated using simulated body fluid (SBF) and MG-63 cells culture experiments, respectively.

RESULTS AND DISCUSSION

Microstructural Analysis

The micromorphologies of the initial powders were displayed in Figure 1. The HAp powders showed a granule shape with an average size of 20 nm (Figure 1a). The CAS powders have a particle size range of 0.5–5 μm (Figure 1b).

Figure 1: The initial powders: (a) hydroxyapatite (HAp); and (b) CaO-Al$_2$O$_3$-SiO$_2$ (CAS).

The micrographs of the HAp ceramics without or with CAS were shown in Figure 2a–f, corresponding to CAS contents of 0, 1, 2, 3, 4 and 5 wt%, respectively. The grains of HAp ceramics without CAS were loose (Figure 2a), which indicated that they were not well sintered. The intergranular spacing of HAp ceramics decreased with increasing CAS from 1 to 2 wt% (Figure 2b,c). When CAS content was 3 wt%, a fully dense structure was obtained, and grain boundary phase appeared (Figure 2d). EDS analysis further demonstrated that grains were consist of Ca, P and O elements (Figure 2g) and grain boundaries were consist of Ca, Si, Al and O elements (Figure 2h), indicating CAS formed a liquid phase around the HAp grains. Moreover, the grains uniformity of the HAp ceramics with 3 wt% CAS significantly improved compared with that of other compositions. The excessive grain boundary phase formed when CAS content was 4 or 5 wt% (Figure 2e,f). It was evident to illustrate that CAS liquid phase could wet HAp crystalline grain and promote densification. The grains shape transformed from polygon to round shape gradually, which indicated the HAp grains dissolved and transformed in sintering process.

It was obvious that CAS could form a viscous flow under the sintering conditions. In liquid phase sintering theory [23], the solid particles were completely wetted by the viscous flow, which generated abundant capillary force between the particles. It facilitated densification via rearranging the particles to produce better arrangement, and via providing a rapid spread channel between the particles to realize rapid mass transfer in dissolution-reprecipitation process. According to the micrographs, though there was no evidence of reprecipitation, CAS formed a viscous flow during sintering.

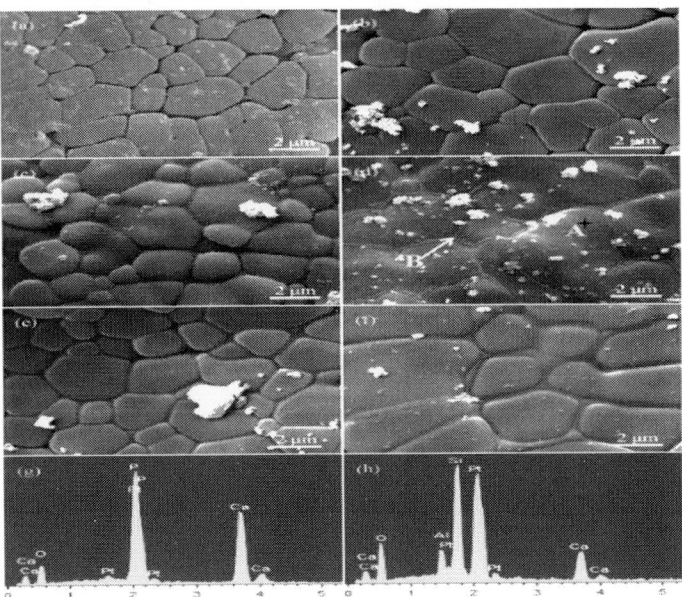

Figure 2: The morphologies of the HAp ceramics without or with CAS: (a) 0; (b) 1; (c) 2; (d) 3; (e) 4; and (f) 5 wt% and energy dispersive spectroscopy (EDS) spectrums: (g) A; (h) B.

Mechanical Characterization

The effects of CAS content on the compression strength and fracture toughness of the HAp ceramics were illustrated in Figure 3. According to the histogram, both the compression strength and fracture toughness plots showed a similar trend. The compression strength and fracture toughness were enhanced when CAS was added. The optimal compression strength and fracture toughness were achieved for the HAp ceramics with 3 wt% CAS, reaching 22.22 ± 0.77 MPa and 1.68 ± 0.06 MPa·m$^{1/2}$, respectively, which represented 105% improvement in compression strength and 63% improvement in fracture toughness compared with those of the HAp ceramics without CAS. However, both of the compression strength and fracture toughness were decreased with further increasing CAS.

The results indicated that CAS had a positive effect on the compression strength and fracture toughness. The improvement of compression strength and fracture toughness was attributed to uniform grains and improved densification, which was achieved through the solid particle rearrangement caused by capillary force during liquid phase sintering [32]. The more additive doped, the better densification was. Moreover, if the ceramics were sintered to nearly full density, the presence of CAS decreased the mechanical properties due to the excessive grain boundary phase, which separated HAp grains [33]. The HAp ceramics with 3 wt% CAS exhibited favorable mechanical properties; so, they were chosen to perform *in vitro* bioactivity and biocompatibility tests.

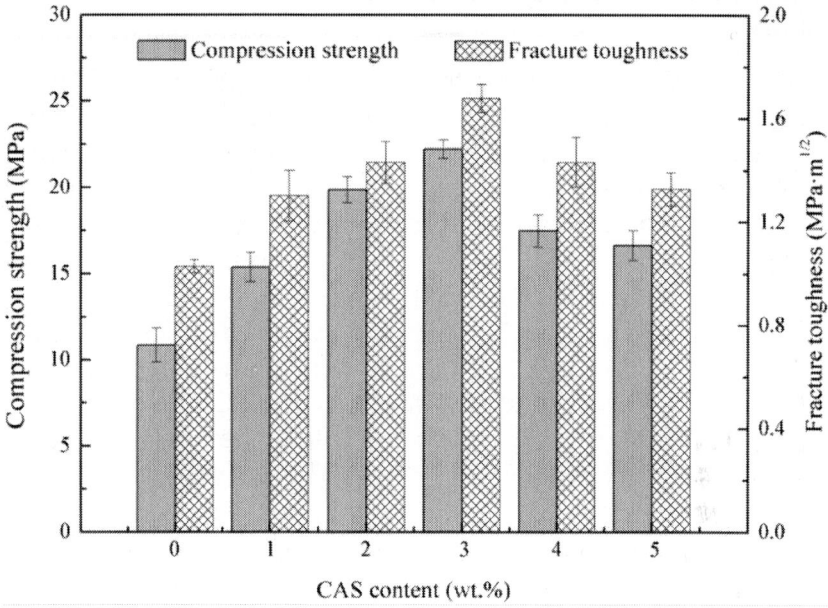

Figure 3: Compression strength and fracture toughness of the HAp ceramics as a function of CAS.

SBF Study

The average Vickers hardness of the HAp ceramics without or with CAS was shown in Figure 4. The Vickers hardness was increased with increasing CAS from 0 to 3 wt%. However, it decreased with further increasing CAS to 4 or 5 wt%. The maximum Vickers hardness of 4.47 ± 0.05 GPa (an increase of 11% over that of the HAp ceramics without CAS) was obtained with 3 wt% CAS. The results showed a similar variation with the compression strength and fracture toughness. After immersion in SBF at different time, the Vickers hardness was decreased slightly, which indicated that the HAp ceramics could maintain the mechanical properties even if they were immersed in SBF. In general, to maintain mechanical properties during immersion period, it was necessary to develop the bone scaffolds with controlled mechanical properties loss [34].

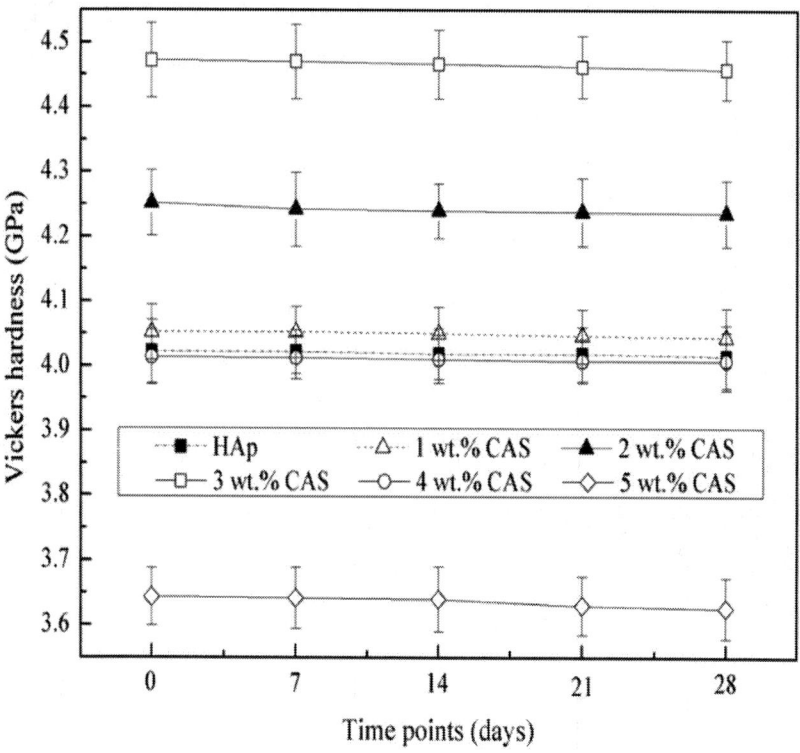

Figure 4: Vickers hardness of the HAp ceramics as a function of immersion time in simulated body fluid (SBF).

The morphologies of apatite on the HAp ceramics with 3 wt% CAS after immersion in SBF at different time were observed by SEM (Figure 5). At initial stage of immersion, some precipitations formed and dispersed on the HAp ceramics surface (Figure 5a). After immersion for 14 days, the precipitation nucleated whilst apatite almost covered on the HAp ceramics surface completely (Figure 5b). At day 21 after soaking in SBF, the porous hemispherical pellets were observed (Figure 5c). After 28 days, these hemispherical pellets were further connected to each other and formed a sponge-like apatite layer (Figure 5d). The bone-like apatite layer formed as consequence of the dissolution and precipitation process of the HAp ceramics [35]. Besides, Ca, P and Si ions were released from the HAp ceramics and presented nucleus to form the apatite. The results of SBF study indicated that the HAp ceramics with 3 wt% CAS had an outstanding bioactivity.

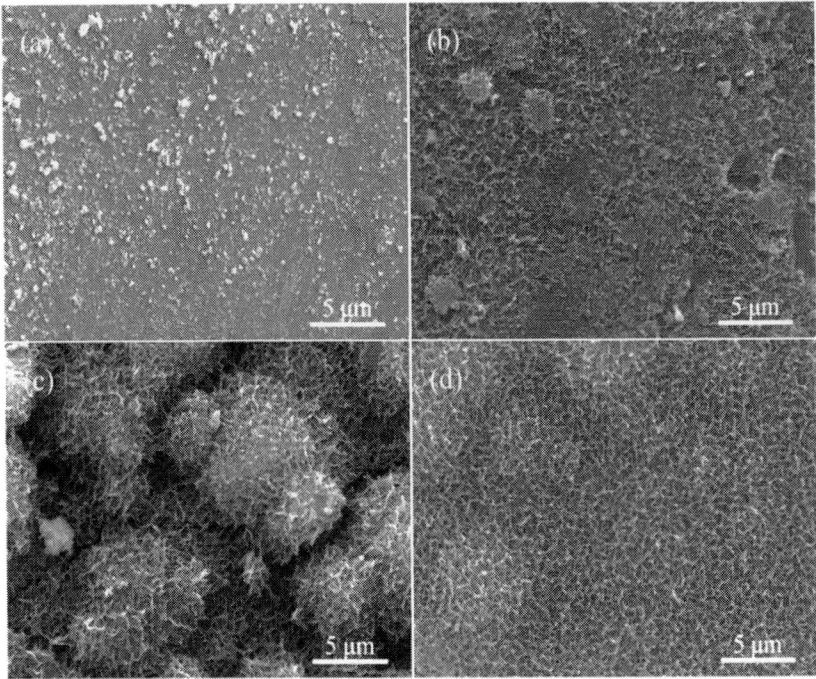

Figure 5: The HAp ceramics with 3 wt% CAS after immersion in SBF for (a) 7; (b) 14; (c) 21; and (d) 28 days.

Cell Culture

Osteoblast-like MG-63 cells were cultured on the HAp ceramics for 1, 3 and 5 days, respectively. The micromorphological features of cells after cultivation were shown in Figure 6. At day 1 after seeding, the MG-63 cells adhered with pseudopodia on the HAp ceramics, and exhibited a spherical morphology (Figure 6a). After 3 days, the cells changed to flat morphology. Besides, extracellular matrix, which was secreted by the MG-63 cells, appeared and covered on the HAp ceramics (Figure 6b). With the culture time increasing to 5 days, the HAp ceramics were completely covered with the cells and extracellular matrix (Figure 6c). Simultaneously, the cells exhibited elongated and flat morphology, and proliferated well on the HAp ceramics. This *in vitro* experiment indicated that the HAp ceramics with 3 wt% CAS possessed excellent biocompatibility in terms of the MG-63 cells culturing.

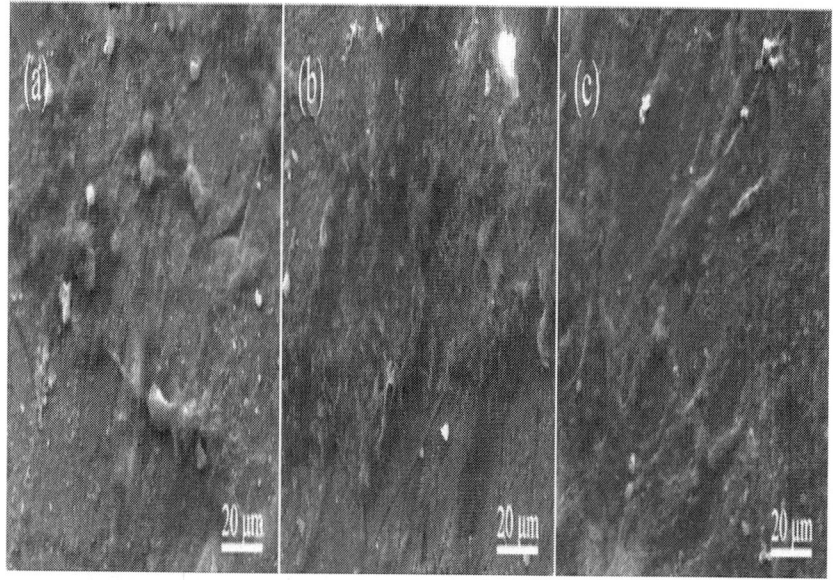

Figure 6: The MG-63 cells cultured on the HAp ceramics with 3 wt% CAS for (a) 1; (b) 3; and (c) 5 days.

EXPERIMENTAL SECTION

Materials and Method

HAp composite powders with different amounts of CAS were prepared in following steps. Firstly, CAS was prepared by mixing 23.3 wt% CaO (Kemiou Chemical Reagent Co., Tianjin, China), 14.7 wt% Al_2O_3 (Wanjing New Material Co., Hangzhou, China) and 62.0 wt% SiO_2 (Emperor Nano Material Co., Nanjing, China) in an appropriate amount of anhydrous alcohol followed by ultrasonic dispersion for 50 min. The dispersed powders were put in the drying oven at 60 °C for 8 h. Afterwards, they were ground into fine powders by using a mortar and pestle. Commercially available medical grade HAp (Emperor Nano Material Co., Nanjing, China) was used for this research. Finally, 1, 2, 3, 4 and 5 wt% of CAS were added into HAp, respectively, followed by ball milling for 6 h.

A home-made SLS system was used in this study [36]. In the SLS system, the scaffolds preparation was conducted using the following procedure: firstly, the prepared HAp composite powders with a layer thickness of 0.1–0.2 mm were laid on working platform. Secondly, the powders were sintered selectively using a focused laser beam. The sintering track was controlled directly by computer-aided design (CAD). Thirdly, the working platform moved down with a layer's height while the sintering of former layer was completed. Subsequently, the powders were laid and sintered repeatedly. Finally, when sintering was completed, three-dimensional scaffolds were obtained after brushing off the unsintered powders externally and internally. All sintering parameters were remained constant during sintering and shown in Table 1. The scheme of the preparation of CAS-containing HAp was presented in Figure 7.

Table 1: Parameters setting for selective laser sintering

Parameters	Spot Diameter (mm)	Scan Spacing (mm)	Laser Power (W)	Scan Speed (mm/min)	Layer Thickness (mm)
Value	1.0	2.0	6	100	0.1–0.2

Microstructural Analysis

The HAp ceramics without or with CAS were etched using 5% hydrofluoric acid solution for 5 min. The initial powders and HAp ceramics were coated with gold for 100 s in a sputtering machine (JFC-1600 auto fine coater, JEOL Ltd., Tokyo, Japan). Scanning electron microscopy (SEM, TESCAN MIRA3 LMU, Co., Brno, Czech) was used to observe and analyze the microstructure of the initial powders and HAp ceramics. Energy dispersive spectroscopy (EDS, Oxford X-Max20, Inc., Oxford, UK) was performed for chemical microanalysis.

Mechanical Characterization

The compression strength, fracture toughness and Vickers hardness of the HAp ceramics with different contents of CAS were investigated.

The compression strength was tested by a universal testing machine (Zhuoji Instruments Co., Ltd., Shanghai, China) with the crosshead speed of 0.5 mm/min. The compression strength was determined by the stress-strain curve obtained from the compression test.

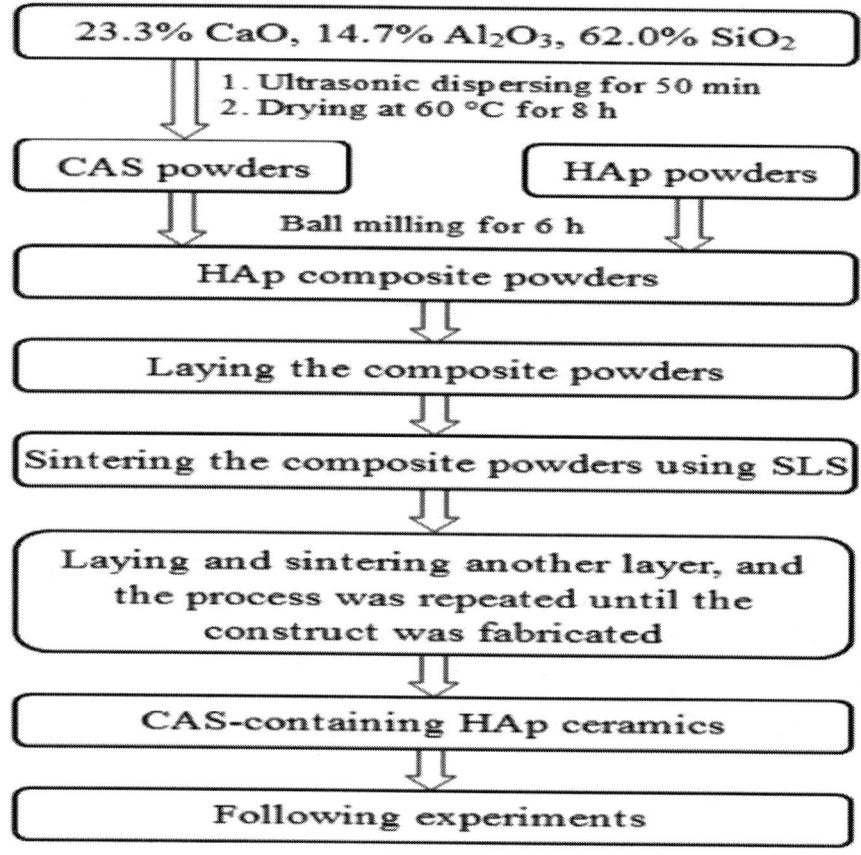

Figure 7: Scheme for the preparation of CAS-containing HAp.

The Vickers hardness was tested on the polished surface of the HAp ceramics using a Vickers microhardness tester (HXD-1000TM/LCD, Digital Micro Hardness Tester, Taiming Optical Instrument Co., Ltd., Shanghai, China) with a maximum load of 4.9 N and an interaction time of 15 s. In the test, the indentation was obtained with a Vickers diamond indenter. The fracture toughness was calculated by the Equation (1) [37]:

$$K_{IC} = 0.0824(P/c^{3/2})$$

(1)

where K_{IC} is the fracture toughness (MPa·m$^{1/2}$); P is the indentation load (N); and c is the induced radial crack length (m). In this study, five specimens with each composition were performed to get an average compression strength, fracture toughness and Vickers hardness. The results were recorded as means ± standard deviation.

SBF Study

SBF has been used widely to study the *in vitro* bioactivity of bioceramic materials by examining the formation of bone-like apatite layer, which played an important impact on the tissue adhesion [38]. In this study, SBF solution was prepared according to the process introduced by Kokubo *et al.* [39]. The ion concentrations of SBF were similar to human body plasma as shown in Table 2. The HAp ceramics were immersed in SBF solution, which was replaced every day. The HAp ceramics were extracted from SBF after 7, 14, 21 and 28 days, respectively, followed by cleaning out with distilled water and dried at 60 °C. After immersion in SBF, the Vickers hardness of the HAp ceramics was measured; the morphologies of the HAp ceramics were observed under SEM. .

Table 2: The ions concentration of human blood plasma and SBF

Ions	Ions Concentration (mmol·L−1)							
	Na+	K+	Ca2+	Mg2+	Cl−	HPO42−	HCO3−	SO42−
Blood plasma	142.0	5.0	2.5	1.5	103.0	1.0	27.0	0.5
SBF	142.0	5.0	2.5	1.5	147.8	1.0	4.2	0.5

Cell Culture

Osteoblast-like MG-63 cells were used to investigate the biocompatibility of the HAp ceramics. Before seeding, the HAp ceramics were sterilized by 70% ethanol for 30 min and then further sterilized by ultraviolet light (UV) for 1 h. Cells were seeded on the HAp ceramics at a density of 3×10^4 cells/well in a 12-well plate. Cultivation was performed in Dulbecco's Modified Eagle's Medium (DMEM) supplementing with 10% fetal bovine serum (FBS), and maintained in a humidified condition containing 5% CO_2 at 37 °C for 1, 3 and 5 days, respectively. After every culture time, the HAp ceramics were extracted from the DMEM and washed with phosphate buffered saline (PBS) to remove non-adherent cells. The cells were fixed in 2.5% glutaraldehyde for 40 min and dehydrated in an alcohol concentration gradient (70%, 80%, 90% and 100%) for 10 min, respectively. The morphologies of MG-63 cells after different culture time were observed under SEM. The scheme of cultivation of MG-63 cells was presented in Figure 8.

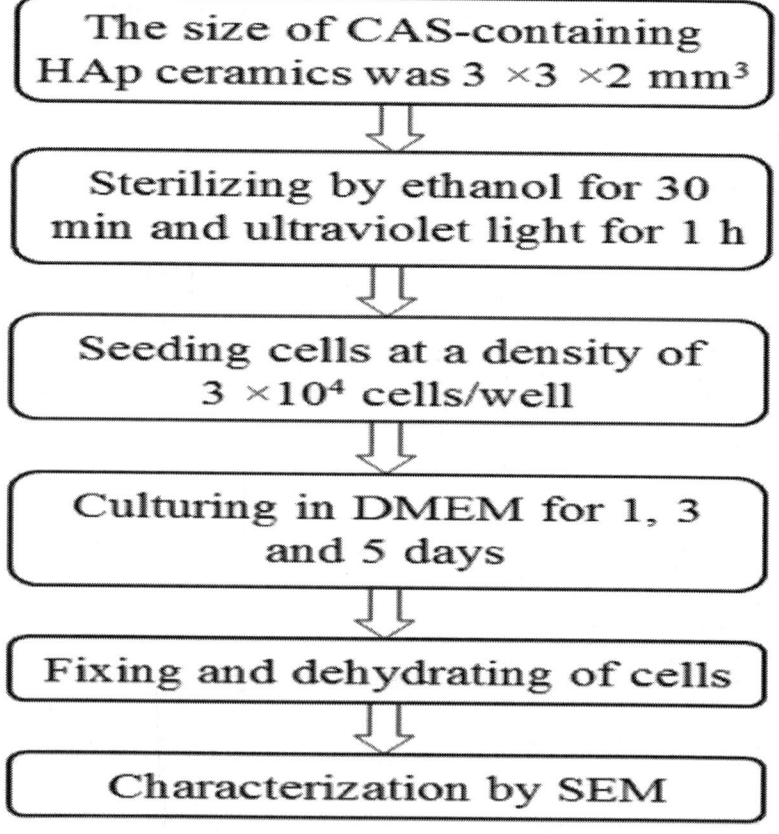

Figure 8: Scheme for cultivation of MG-63 cells.

CONCLUSIONS

In this paper, the HAp ceramics with different contents of CAS (0, 1, 2, 3, 4 and 5 wt%) were prepared by SLS. Mechanical properties of the HAp ceramics were improved significantly by adding a small quantity of CAS as a liquid phase. The mechanical properties with 105% increase in compression strength, 63% increase in fracture toughness and 11% increase in Vickers hardness were obtained. The improvement of mechanical properties was attributed to the improved densification. However, mechanical properties were decreased, while 4 or 5 wt% CAS was added, which was due to the excessive liquid phase. Moreover, *in vitro* bioactivity study carried out in SBF solutions showed the formation of bone-like apatite layer on the HAp ceramics. Meanwhile, the Vickers hardness of the HAp ceramics was decreased slightly after immersion in SBF, indicating excellent mechanical stability. In addition, cell culture was performed with the MG-63 cells, and the results indicated that the cells adhered and spread well. Therefore, the HAp ceramics with CAS were bioactive and biocompatible substitutes, which might be suitable for bone repair.

ACKNOWLEDGMENTS

This work was supported by the following funds: (1) The Natural Science Foundation of China (51222506, 81372366, 81472058); (2) High Technology Research and Development Program of China (2015AA033503); (3) Overseas, Hong Kong & Macao Scholars Collaborated Researching Fund of National Natural Science Foundation of China (81428018); (4) Hunan Provincial Natural Science Foundation of China (14JJ1006); (5) Shenzhen Strategic Emerging Industrial Development Funds (JCYJ20130401160614372); (6) The Open-End Fund for the Valuable and Precision Instruments of Central South University; (7) The faculty research grant of Central South University (2013JSJJ011, 2013JSJJ046); (8) State Key Laboratory of New Ceramic and Fine Processing Tsinghua University (KF201413).

AUTHOR CONTRIBUTIONS

Songlin Duan, Pei Feng and Chengde Gao performed ceramics preparation and took part in the mechanical testing of the ceramics under the supervision of Cijun Shuai, the mechanical characterization of the ceramics under the supervision of Cijun Shuai and Kun Yu, the biological testing of the ceramics under the supervision of Shuping Peng and the biological characterization of the ceramics under the supervision of Shuping Peng and Tao Xiao. All authors took part in the discussion of results and the preparation of the manuscript.

REFERENCES

1.   Soucacos, P.N.; Kokkalis, Z.T.; Piagkou, M.; Johnsonb, E.O. Vascularized bone grafts for the management of skeletal defects in orthopaedic trauma and reconstructive surgery. *Injury* 2013, *44*, S70–S75.

2.   Schlickewei, W.; Schlickewei, C. The use of bone substitutes in the treatment of bone defects–The clinical view and history. *Macromol. Symp. Aug.* 2007, *253*, 10–23.

3.   Shi, Y.; Niedzinski, J.R.; Samaniego, A.; Bogdansky, S.; Atkinson, B.L. Adipose-derived stem cells combined with a demineralized cancellous bone substrate for bone regeneration. *Tissue Eng. Part A* 2012, *18*, 1313–1321.

4.   Goodridge, R.D.; Dalgarno, K.W.; Wood, D.J. Indirect selective laser sintering of an apatite-mullite glass-ceramic for potential use in bone replacement applications. *Proc. Inst. Mech. Eng. H* 2006, *220*, 57–68.

5.   Kaur, G.; Pandey, O.P.; Singh, K.; Homa, D.; Scott, B.; Pickrell, G. A review of bioactive glasses: Their structure, properties, fabrication and apatite formation. *J. Biomed. Mater. Res. A* 2014, *102*, 254–274.

6.   Martínez-Vázquez, F.J.; Pajares, A.; Guiberteau, F.; Miranda, P. Effect of polymer infiltration on the flexural behavior of β-tricalcium phosphate robocast scaffolds. *Materials* 2014, *7*, 4001–4018.

7.   Lawrence, B.J.; Madihally, S.V. Cell colonization in degradable 3D porous matrices. *Cell Adhes. Migr.*2008, *2*, 9–16.

8.   Guan, S.; Zhang, X.L.; Lin, X.M.; Liu, T.Q.; Ma, X.H.; Cui, Z.F. Chitosan/gelatin porous scaffolds containing hyaluronic acid and heparan sulfate for neural tissue engineering. *J. Biomater. Sci.-Polym. Ed.* 2013, *24*, 999–1014.

9.   Murphy, C.M.; Haugh, M.G.; O'Brien, F.J. The effect of mean pore size on cell attachment, proliferation and migration in collagen–glycosaminoglycan scaffolds for bone tissue engineering. *Biomaterials* 2010, *31*, 461–466.

10. Zhou, H.; Lee, J. Nanoscale hydroxyapatite particles for bone tissue engineering. *Acta Biomater.* 2011,*7*, 2769–2781.

11. Sun, F.; Koh, K.; Ryu, S.C.; Han, D.W.; Lee, J. Biocompatibility of nanoscale hydroxyapatite-embedded chitosan films. *Bull. Korean. Chem. Soc.* 2012, *33*, 3950–3956.

12. Yang, Y.H.; Liu, C.H.; Liang, Y.H.; Lin, F.H.; Wu, K.C.W. Hollow mesoporous hydroxyapatite nanoparticles (hmHANPs) with enhanced drug loading and pH-responsive release properties for intracellular drug delivery. *J. Mater. Chem. B* 2013, *1*, 2447–2450.

13. Wu, K.C.W.; Yang, Y.H.; Liang, Y.H.; Chen, H.Y.; Sung, E.; Yamauchi, Y.; Lin, F.H. Facile synthesis of hollow mesoporous hydroxyapatite nanoparticles for intracellular bio-imaging. *Curr. Nanosci.* 2011,*7*, 926–931.

14. Li, Z.; Wen, T.; Su, Y.; Wei, X.; He, C.; Wang, D. Hollow hydroxyapatite spheres fabrication with three-dimensional hydrogel template. *Cryst. Eng. Comm.* 2014, *16*, 4202–4209.

15. Bastakoti, B.P.; Hsu, Y.C.; Liao, S.H.; Wu, K.C.W.; Inoue, M.; Yusa, S.I.; Yamauchi, Y. Inorganic–organic hybrid nanoparticles with biocompatible calcium phosphate thin shells for fluorescence enhancement. *Chem. Asian J.* 2013, *8*, 1301–1305.

16. Chevalier, J.; Gremillard, L. Ceramics for medical applications: A picture for the next 20 years. *J. Eur. Ceram. Soc.* 2009, *29*, 1245–1255.

17. Lin, K.; Chen, L.; Chang, J. Fabrication of dense hydroxyapatite nanobioceramics with enhanced mechanical properties via two-step sintering process. *Int. J. Appl. Ceram. Technol.* 2012, *9*, 479–485.

18. Reves, B.T.; Jennings, J.A.; Bumgardner, J.D.; Haggard, W.O. Osteoinductivity assessment of BMP-2 loaded composite chitosan-nano-hydroxyapatite scaffolds in a rat muscle pouch. *Materials* 2011, *4*, 1360–1374.

19. Tripathi, G.; Basu, B. A porous hydroxyapatite scaffold for bone tissue engineering: Physico-mechanical and biological evaluations. *Ceram. Int.* 2012, *38*, 341–349.

20. Duan, B.; Wang, M. Selective laser sintering and its application in biomedical engineering. *MRS Bull.* 2011, *36*, 998–1005.

21. Sandler, N.; Lammens, R.F. Pneumatic dry granulation: Potential to improve roller compaction technology in drug manufacture. *Expert Opin. Drug Del.* 2011, *8*, 225–236.

22. Bouslama, N.; Chevalier, Y.; Bouaziz, J.; Ayed, F.B. Influence of the sintering temperature on Young's modulus and the shear modulus of tricalcium phosphate–fluorapatite composites evaluated by ultrasound techniques. *Mater. Chem. Phys.* 2013, *141*, 289–297.

23. Lupulescu, A.; Glicksman, M.E. Diffusion-limited crystal growth in silicate systems: Similarity with high-pressure liquid-phase sintering. *J. Cryst. Growth* 2000, *211*, 49–61.

24. Wei, W.; Chen, K.; Ge, G. Strongly coupled nanorod vertical arrays for plasmonic sensing. *Adv. Mater.* 2013, *25*, 3863–3868.

25. Hotta, M.; Hojo, J. Inhibition of grain growth in liquid-phase sintered SiC ceramics by AlN additive and spark plasma sintering. *J. Eur. Ceram. Soc.* 2010, *30*, 2117–2122.

26. Liu, D.; Zhuang, J.; Shuai, C.; Peng, S. Mechanical properties' improvement of a tricalcium phosphate scaffold with poly-L-lactic acid in selective laser sintering. *Biofabrication* 2013, *5*, 025005.

27. Batra, U.; Kapoor, S.; Sharma, J.D. Nano-Hydroxyapatite/ Fluoridated and Unfluoridated Bioactive Glass Composites: Structural Analysis and Bioactivity Evaluation. In Proceedings of the International Conference on Advances in Condensed and Nano Materials (ICACNM-2011), Chandigarh, India, 23–26 February 2011; AIP Publishing: New York, NY, USA, 2011; Volume 1393, pp. 271–272.

28. Oktar, F.N.; Agathopoulos, S.; Ozyegin, L.S.; Gunduz, O.; Demirkol, N.; Bozkurt, Y.; Salman, S. Mechanical properties of

bovine hydroxyapatite (BHA) composites doped with SiO_2, MgO, Al_2O_3, and ZrO_2. *J. Mater. Sci.: Mater. Med.* 2007, *18,* 2137–2143.

29. Bellucci, D.; Cannillo, V.; Sola, A. A new highly bioactive composite for scaffold applications: A feasibility study. *Materials* 2011, *4,* 339–354.

30. Nath, S.; Biswas, K.; Wang, K.; Bordia, R.K.; Basu, B. Sintering, phase stability, and properties of calcium phosphate-mullite composites. *J. Am. Ceram. Soc.* 2010, *93,* 1639–1649.

31. Bose, S.; Tarafder, S.; Banerjee, S.S.; Davies, N.M.; Bandyopadhyay, A. Understanding *in vivo* response and mechanical property variation in MgO, SrO and SiO_2 doped β-TCP. *Bone* 2011, *48,* 1282–1290.

32. German, R.M.; Suri, P.; Park, S.J. Review: Liquid phase sintering. *J. Mater. Sci.* 2009, *44,* 1–39.

33. Borrero-López, O.; Ortiz, A.L.; Guiberteau, F.; Padture, N.P. Effect of liquid-phase content on the contact-mechanical properties of liquid-phase-sintered α-SiC. *J. Eur. Ceram. Soc.* 2007, *27,* 2521–2527.

34. Bhatt, H.A.; Kalita, S.J. Influence of oxide-based sintering additives on densification and mechanical behavior of tricalcium phosphate (TCP). *J. Mater. Sci.: Mater. Med.* 2007, *18,* 883–893.

35. Ribeiro, C.; Rigo, E.C.S.; Sepúlveda, P.; Bressiani, J.C.; Bressiani, A.H.A. Formation of calcium phosphate layer on ceramics with different reactivities. *Mater. Sci. Eng. C* 2004, *24,* 631–636.

36. Shuai, C.J.; Feng, P.; Gao, C.D.; Zhou, Y.; Peng, S.P. Simulation of temperature field during the laser sintering process of nano-hydroxyapatite powder. *Adv. Mater. Res.* 2011, *314,* 626–629.

37. Veljovic, D.; Palcevskis, E.; Zalite, I.; Petrovic, R.; Janackovic, D. Two-step microwave sintering-A promising technique for the processing of nanostructured bioceramics. *Mater. Lett.* 2013, *93,* 251–253.

38. Kaur, G.; Pickrell, G.; Kimsawatde, G.; Homa, D.; Allbee, H.A.; Sriranganathan, N. Synthesis, cytotoxicity, and hydroxyapatite formation in 27-Tris-SBF for sol-gel based CaO-P_2O_5-SiO_2-B_2O_3-ZnO bioactive glasses. *Sci. Rep.* 2014, *4,* 4392.

39. Kokubo, T.; Kushitani, H.; Sakka, S.; Kitsugi, T.; Yamamuro, T. Solutions able to reproduce *in vivo* surface-structure changes in bioactive glass-ceramic A-W[3]. *J. Biomed. Mater. Res.* 1990, *24*, 721–734.

Chapter 2

BIM Based Collaborative and Interactive Design Process Using Computer Game Engine for General End-users

Gareth Edwards[1], Haijiang Li[2], and Bin Wang[2]

[1]ATKINS, The Hub, 500 Park Avenue, Aztec West, Almondsbury, Bristol BS32 4RZ, UK

[2]Cardiff School of Engineering, Cardiff University, Queen's Building, The Parade, Cardiff CF24 3AA, Wales

ABSTRACT

Background

The emerging Building Information Modelling (BIM) in the Architectural, Engineering and Construction (AEC)/Facility Management (FM) industry promotes life cycle process and collaborative way of working.

Currently, many efforts have been contributed for professional integrated design/construction/maintenance process, there are very few practical methods that can enable a professional designer to effectively interact and collaborate with end-users/clients on a functional level.

Method

This paper tries to address the issue via the utilisation of computer game software combined with Building Information Modelling (BIM). Game-engine technology is used due to its intuitive controls, immersive 3D technology and network capabilities that allow for multiple simultaneous users. BIM has been specified due to the growing trend in industry for the adoption of the design method and the 3D nature of the models, which suit a game engine's capabilities.

Results

The prototype system created in this paper is based around a designer creating a structure using BIM and this being transferred into the game engine automatically through a two-way data transferring channel. This model is then used in the game engine across a number of network connected client ends to allow end-users to change/add elements to the design, and those changes will be synchronized back to the original design conducted by the professional designer. The system has been tested for its robustness and functionality against the development requirements, and the results showed promising potential to support more collaborative and interactive design process.

Conclusion

It was concluded that this process of involving the end-user could be very useful in certain circumstances to better elaborate the end user's requirement to design team in real time and in an efficient way.

BACKGROUND

Building Information Modelling (BIM) is a new technology/method emerging in AEC/FM industry. It promotes life cycle process and collaborative way of working. The basic concept is that all information/ data are shared/updated/maintained by all of the involved parties guided/governed by the appropriate governance/management model. BIM allows for design of a building/structure to be completed in full 3D and also allows for a design to be modified and updated for all parties so that they can see the changes almost immediately and see if those changes affect their elements of the design. Additionally, BIM enables more specialised parts of the design to be completed by others, other than the traditional main three (architect, structural engineer and service engineer). For example it facilitates conducting energy analysis or quantity surveyors to look at the quantities of materials that may be required based on the elements parametric data. It also aids scheduling of the works, known as 4-Dimensional (4D) design. This schedule can be used to create an animation of the construction project, which may be useful when helping professional designer to understand the full construction process or demonstrate it to a client. BIM can also allow the non-technical parties to understand the design at the design stage, to see the design in 3D and get a proper visualisation of the whole project, which is not easily possible with the more traditional methods.

Although, in theory BIM does allow for far greater collaboration between many of the parties involved in the design process, the current practices and developments mainly focus on the collaboration between the different elements of the professional design teams. Many efforts have been contributed for professional integrated design / construction / maintenance process, there are very few practical methods that can enable a professional designer to effectively interact and collaborate with end users (functionally). Currently, with either BIM or traditional design practices the client is generally only provided with passive information on a project via the design team preparing information in some form of formal presentation. On a functional level BIM has not yet allowed for a client to contribute to the design phase (Christiansson et al. [2011]). By this it is meant that all of the data that is presented to a client is static and has to be transferred into some sort of other form with no real interactive element for the client. An example of this would be that the model might be presented on paper or on a computer to a client but the views and information will have been pre-prepared by one of the design team to present to the client and may not present all of the information that a client wants to see. This also does not allow the client to actively explore the building in its entirety and view what they want to see not what a designer thinks that they want/need to see.

This paper introduces a software system development addressing the technical issue as mentioned above regarding the involvements of end users in BIM based design process via the utilisation of computer game engine combined with BIM (Christiansson et al. [2011]; Eastman et al. [2008]). A computer game engine is used due to its intuitive controls (easy enough for non-professional end users), immersive 3D technology and network capabilities that allow for multiple simultaneous users. BIM has been specified due to the growing trend in industry for the adoption of the design method and the 3D nature of the models, which suit a game engine's capabilities. The key contribution is to show a way that allows the functional involvement of non-professional end users in the professional design process. The core research is to develop a two-way data transferring channel between the end users 'playing' in game environment (through Web pages or thin client end) and the professional BIM design team. The other supporting developments include the extraction of the necessary information from the professional BIM model (through C# based Revit API development), game environment development, and setting up the underlying computing environment etc. The developed prototype further extends the current BIM implementation to cover non-professional end users. With further improvement and development, e.g., embedded with the appropriate governance model, it could be used to realize a more integrated life cycle based building design/construction/maintenance process to better address the end user's requirements, and hence to improve the efficiency and productivity in the industry.

The rest of the paper is organized as follows. First, the related work regarding computer game engine with BIM applied in AEC/FM domain has been briefly reviewed, including computer game engine software infrastructure and their various applications. A comprehensive software system development process is then explained in detail. The software system architecture is introduced first via a multi-layered diagram, which includes the client end, game engine environment, BIM environment and the data transmission among different elements.

The system design and implementation section covers the main hierarchical system design, key classes design and implementation, development framework selection, the implementation for two-way data transferring, extraction of BIM data and other development details. The system deployment and evaluation is given next, where several testing scenarios have been implemented to check the software development errors and system functionalities. Finally, the conclusion and future work are given.

Game Engine Applied in AEC/FM Domain with BIM

Building Information Modelling (BIM) is now approaching to its tipping point (NBIMS [2007]; BSI [2010]; BuildingSmart [2010]; McGraw_Hill_construction [2010]) worldwide to revolutionize the entire workflow for AEC industry. Like other countries all over the world, the UK has completed several strategic BIM implementation plans, which are going to require BIM compatibility (level 2 of BIM in Bew & Richards Figure (BuildingSmart [2010])) by the year of 2016. There is a drive from the current UK AEC industry to implement the comprehensive and advanced BIM practice to fulfil the government requirement. While computer game engines have been developed over the years from simple games such as the first interactive computer game OXO (Obituary [2011]) to fully immersive 3D environments. Games are highly profitable, thus developers are willing to invest heavily in resources for games development. This gives rise to a highly developed product. The net result is that games represent a pinnacle that other simulation and virtualisation software struggle to equal (Stuart [2011]). All of this leads to very powerful tools that can be used for various purposes.

One of the situations that computer game engine technology is being applied to is that of simulating evacuation of the population of a building in a situation such as a fire (Rüppel & Schatz [2011]). The models used in some of these simulations are based on 3D information taken from BIM applications (Yan et al. [2011]). Some of the models use agent based systems where the computer based players are programmed using parameters gained from research and the computer simulates the virtual players attempting to evacuate the building (Rüppel & Schatz

[2011]). This has been extended and changed from being simply a fully computer driven simulation to one utilising the interactivity of the game engine by human subjects to identify how they react within the simulation and what decisions they make in the stressful situation of a fire. Rüppel and Schatz ([2011]) took this simulation further introducing stimuli such as smoke machines and 3D optical effect screens etc. to try to bring a new dimension to the game.

Another way in which computer game engines have been put to use when combined with BIM is that of providing an interactive visualisation of a structure. This has been done in (Yan et al. [2011]) using a combination of the Microsoft® XNA™ Framework as the game engine and Autodesk® Revit® Architecture as the BIM design application. The system allows users to go on a 'walkthrough' from one room to another or simply explore the building freely as they wish. This allows for a far more intuitive method of exploring a design before it is built or whilst the building is under construction when compared to the static 2D drawings produced by most CAD based systems. The system also makes full use of the parametric property aspect of the BIM design process by using the properties that indicate if rooms are linked by a door etc. to allow for path finding in the building. A further use that game engines appear to suit is one of visualising a simulation in a graphical way. They suit themselves to this situation well due to their real-time, high quality 3D graphics capabilities and also their ability to cope with 3D geometric data.

El Nimr and Mohamed ([2011]) used two different scenarios and two different game engines to achieve visualisations of the simulations that they were running. The first example was that of a system designed to simulate the bidding process for attaining a project from tender. It used a map on which the simulation created jobs that players could then bid for. The second example used a real-life construction scenario and a system was developed to visualise the construction process on a site where parts of the structure were to be built off-site. These were brought to a yard on site for further assembly into modules that were then finally constructed into the finished structure. This system simulated the whole construction process and visualised both the site, enabling the contractors to visualise where cranes could fit on to the site during the construction, and also the yard where the individual parts were to be assembled into modules. Both of the scenarios demonstrated the merit of having a graphical representation of the simulation, which allowed for the developers and also the users of the systems to see what was going on and where that fitted into the project.

Using a computer game engine as part of a design review system may not seem the obvious choice. However, due to the nature of the game engine providing networking features that enable real-time collaboration and real-time 3D graphics for real-time visualisation of a building, a game engine could perform this task well. Currently many of the design review systems that are available provide insufficient allowance for the inclusion of all of the stakeholders who may sensibly contribute to this process. Current methods generally are conducted using peer review. An example of this is the use of a light box to overlay drawings to facilitate discrepancies to be identified between different elements of the design. Due to this sort of process becoming tedious it is found that the process is quite often not conducted efficiently and in some cases is not conducted at all (Shiratuddin & Thabet [2011]). A prototype has been created by (Shiratuddin & Thabet [2011]), where a game engine has been used to create a game where multiple users (members of the design team) can review a design together and interact with the design. They can see details about the design and do basic 3D editing. It is also possible to communicate with one another within the game and leave notes on the design. The authors envisaged that this sort of design review system will help to improve the quality of the design review that it is undertaken and make it a simpler, easier and more satisfying process.

Another area that game engines have been used for in AEC is that of teaching students about construction site safety (Lin et al. [2011]; Dickinson et al. [2011]). In work by Lin et al., a game was created that simulates a construction site. The user walks around the site looking for errors or issues with health and safety. When an issue is identified the user selects the issue and then chooses what is wrong with the situation using the games interface. It was noted that, from the students who were test subjects in the work, the system made the learning process more interactive and enjoyable and that it helped them to memorise the health and safety issues. In addition, such virtual environment can enable end users to learn various design skills (Sun et al. [2014]; GU et al. [2009]). Sun et al. developed a 3D virtual space towards a synchronous distributed design meeting system to allow end-users to sketch or make annotations and have discussions as well as add viewpoints to designs. Gu and Nakapan discussed the benefits and shortcomings of virtual worlds for collaborative design learning and education.

In short, computer game engine has becoming much more advanced and mature (than it was before) to be utilized for AEC/FM industry. In its essence, BIM is life cycle based collaborative process, which would be much more effective if end user gets involved in an efficient manner right from the beginning stage through to building's demolition. Due to the prompt and direct involvements of end user/client (on a functional level), with good governance model, the way of working can further facilitate the prompt response coming from professional designers, and to some extent it can also mitigate the suffer for designer to repeatedly correct errors due to the lack of effective communication between the end user/client and designer. The development to be introduced in this paper would serve as a good add-on to other developments focusing more on professional teams (Eastman et al.[2008]; Gu et al. [2010]; Hassanien Serror et al. [2008]).

METHODS

System Design and Implementation

The use of game engines provides new and exciting opportunities for technologies such as BIM. However, this sort of utilisation is still in its infancy. Currently BIM does not provide a method to involve the end-users of the structures and promote collaboration between them and the design team. This paper proposes to address this issue by trying to promote collaboration between the design teams and the end-users. The new developed system has to be tested for its robustness and functionality against the development requirements, and the results showed promising potential to support more collaborative and interactive design process between professionals and end-users.

In order to construct a BIM based collaborative design environment, the following steps have been identified and concluded to guide the design of the system, and the multi-layered system architecture is showed in Figure 1.

Identification of the program suite the designer will use to create their designs

Establish how the designs will be conveyed to the end-user

Ascertain how to enable the user to have a simple and intuitive process to modify designs

Determination of a method to transfer the information relating to these changes to the designer

Identification of the modules, data sources and executables that are required to support the task.

Plan the order and structure of the implementation.

Plan the methods used for evaluation and testing of the system.

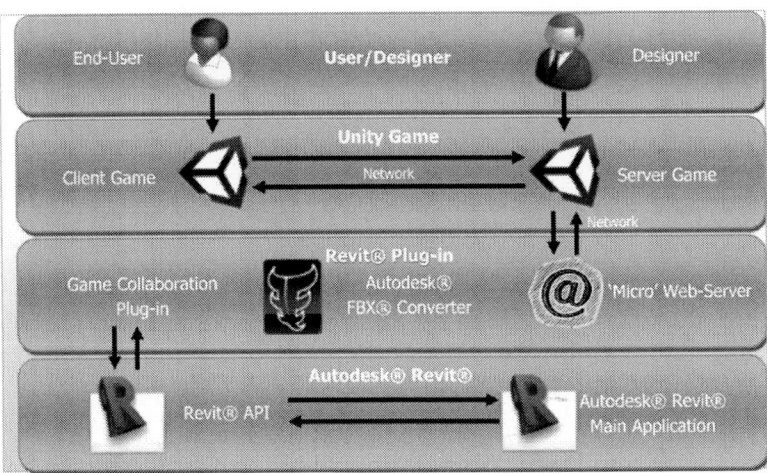

Figure 1: system overall architecture.

The multi-layered system architecture comprises four inter-connected layers (Wang et al. [2014]):

BIM Environment: Autodesk Revit can be regarded as a comprehensive building information provider to work with game technology to build an adaptable immersive serious game

The Data Transmission Element: Revit plug-in contains the database and FBX format converter. It works as 'micro' web-server to generate semantic and geometric data and store it in the database to build a two-way and dynamic information flow for collaborative and interactive design.

Game Engine Environment: Unity game includes the Unity server and client game. A unity server component connects Unity clients into an instance and feeds those clients the available data which is created in the data component. Unity clients are the interface that helps immerse the end-user into the virtual environment which is generated by the Unity server (controlled by the designer).

The Client End: End-users and designers can work together in different platforms supported by Unity game engine with appropriate input and output devices, i.e. Windows or Mac operating systems that use high-resolution monitors with keyboard and mouse; 3D stereoscopic projector with Razer Hydra joystick; Hand Mounted Display (HMD) with Microsoft Kinect Sensor; Mobile platform using iOS or Android with touch screen and built-in camera; or Web-based environment that allows users to connect to the server through their web browser.

Implementation Framework Selection

The major consideration in determining the choice of system development framework was that of compatibility of many of the BIM standards that have been laid down. A further significant consideration was one of third party support with software development kits. Taking these factors into account, Autodesk® Revit® Architecture was selected being fully compatible with BIM standards and having excellent third party developer support (AEC (UK) BIM Standard for Autodesk Revit ([2010])). It should be noted that the system would also function with Autodesk® Revit® Structure and Autodesk® Revit® MEP (Revit 2012 API [2011]) (when working with different professionals). The method of enabling the end-user/client to be presented with and modify information contained within the BIM model must meet the following criteria. The method must be able to display the 3D nature of the model and provide a simple, intuitive interface. This method must also provide some way to facilitate the end-user/client to collaborate with other stakeholders. A suitable system that has been decided upon is that of a game engine. A game engine provides powerful real-time 3D graphics, networking features that allow for collaboration and a simple and intuitive interface (Moore [2011]; Petri et al. [2014]). Unity 3D from Unity Technologies has been chosen as the platform to interact between the professional designer and the end user/client. This particular engine has been chosen due to its simple object orientated and editor based design system. It also provides many features that other engines do not such as the ability to create executables that will run on both Microsoft® Windows® and Apple® Mac®. It also provides the ability to create 'Web player' versions of any game produced, which will be a useful tool when providing a method of simple delivery to the end user as it would mean that the end user could use their own web browsers without having to download and install the game. Currently to use the Unity web player a plug-in must be installed into the web browser being used. This may be negated by using the support for creating an Adobe® Flash® Player object in the 3.5 version of Unity that has been released recently (at the time of writing). Adobe® Flash® Player is a common web technology is utilised on almost all modern websites.

Key Classes Design

All the classes developed in the system help to deliver the collaborative design environment. Figure 2shows the work flow that occurs within the game.

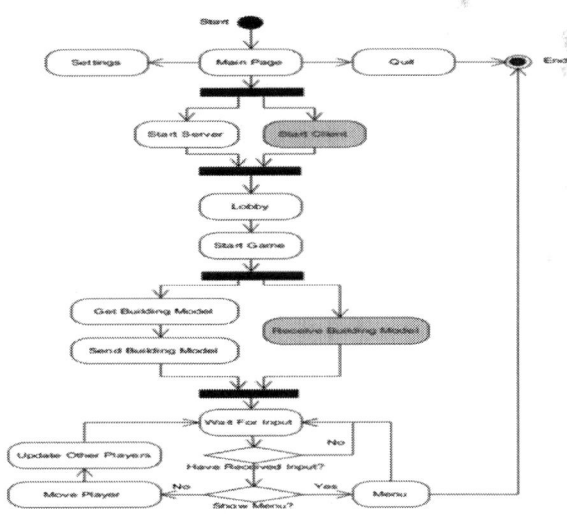

Figure 2: Activity UML model for game flow.

When the server starts up, the first procedure to implement is to create an instance in the database to store this information via the menu system. Next it waits for the building information to load. The loading time varies depending on the amount of information in the BIM model. The internal process of loading data is to convert Revit building format to FBX format. Then the FBX model is loaded into the Unity server's memory and converted into our custom format, which buffers at the network layer for incoming clients to load. The client first needs to connect to the server by entering the server IP address, and then receive the building information. The two highlighted grey sections are specific for the client version of the game and their counterparts are specifically for the server version of the game. The end users can control the avatar and toggle main menu for different tasks in the server and clients.

The entire system development follows a sound software engineering approach, specific activities such as applying metrics and identifying characterisations such as design requirements (aforementioned) and the level of details are important in that activity. In relation to the object oriented software development approach used, the overall system class hierarchy to constitute menu structure and game for collaborative and interactive design is shown in Figures 3 and 4, and the following contents show the key classes design details.

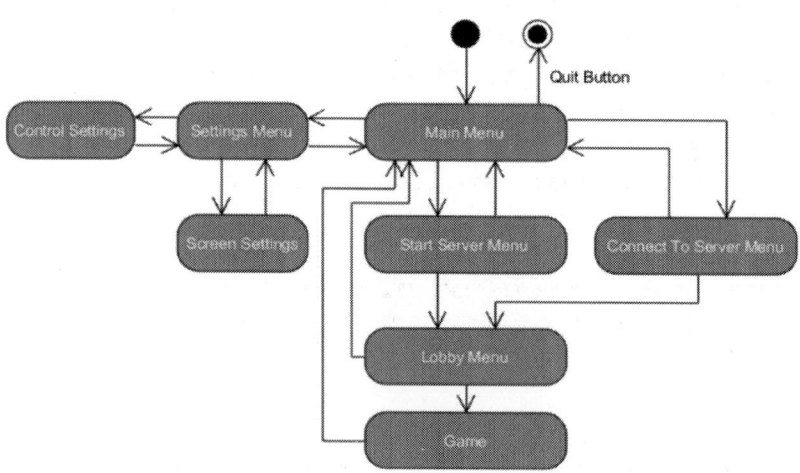

Figure 3: Menu structure of the game.

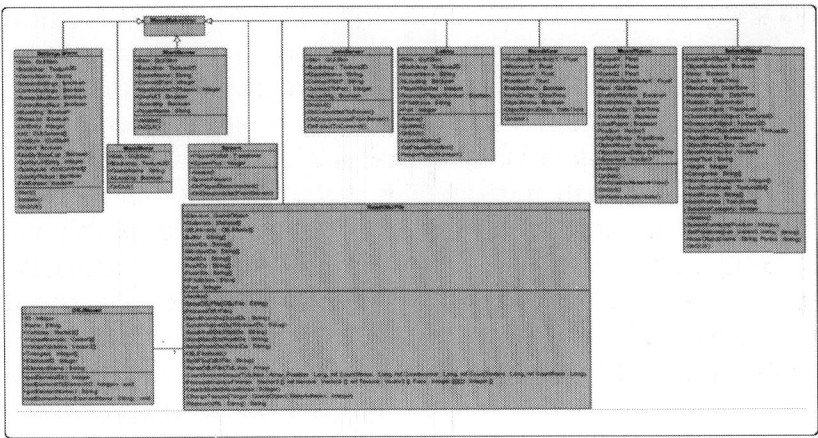

Figure 4: Overall system class hierarchy.

All of the classes except 'OBJModel' are inherited from 'MonoBehaviour', which is Unity's base class.

Main Menu

The 'MainMenu' class which is a single script attached to the camera object in the 'MainMenu' game scene in the Unity game. As can be seen (in Figure 4) it has four variables of which three are public to allow access to them in the Unity editor. These three public variables are a skin for the GUI components to use, a backdrop for the menu and the title or name of the game to display on the menu. These three variables are common on all of the menu scenes as is the fourth variable 'isLoading', which is private and is used for determining if the game is changing scenes between a menu scene or the game scene. Finally it can be seen that there is a subroutine 'OnGUI()' which is a Unity built in subroutine that gets called by the game and is used to create the GUI elements and in this case implement the logic for the buttons of the main menu.

Start Server

The 'StartServer' class, which is attached to the camera object in the 'StartServer' scene. This being a menu has the same variables as the main menu and they will not be described again. It also has some other variables, which are all for the various settings used for the server. All of the variables are all public other than the player number as they need to be accessed by the Unity editor to allow the default settings to be specified. The 'Update()' subroutine, which is built in Unity subroutine that executes once per frame render in the game. It is implemented here to allow for capturing of input keys namely the 'escape' key to get back to the main menu. The 'PlayerName' is included to set a default player preference as part of the network connection starting process. This starting of the game server is initiated in the 'OnGUI()' subroutine along with the other functionality for the controls including moving to the lobby scene.

Connect to Server

The 'JoinServer' class is attached to the main camera in the game scene 'JoinServer'. It also again implements all of the previous standard menu variables and the 'OnGUI()' subroutine. Other than those variables and subroutines it implements two private variables that are used to hold the value of two text input fields for use as settings when trying to connect to a server version of the game. The three other subroutines that are implemented are Unity built in subroutines that as their names suggest deal with various situations of network connectivity and are, in this case, only used to send messages to the debug log. Again the 'OnGUI()' subroutine has been used to implement the functionality of the controls including initiating the connection with the server version of the game and changing to the lobby scene.

Lobby Menu

The 'Lobby' class is attached to the main camera object in the 'Lobby' scene. It implements all of the menu variables and the 'OnGUI()' subroutine. Of the four other variables that are implemented two are used for allocating each player (client) a unique number and the other two are used for allocating the settings that the game server will use to connect the building model server (Revit® plug-in). The 'Update()' subroutine is implemented to capture the 'escape' key input to return to the main menu. The 'Awake()' subroutine is another Unity built in subroutine that initiates before any of the rest of the class starts, unlike the 'Start()' subroutine where some of the class is started before the function runs. It is used to run what is known as a Remote Procedure Call (RPC), which is essentially a method of allowing the same function or subroutine to be initiated on all connect games, in different locations, at the same time. It runs the RPC subroutine 'GetPlayerNumber()', which in turn runs another RPC, namely 'ReturnPlayerNumber' from the clients so that they each request a player number from the server. The functionality of the controls, which are only shown on the game server not the clients, is implemented in the 'OnGUI()' subroutine including executing the 'LaunchGame()' RPC on all clients that makes the server and all clients change scenes to the 'Game' scene.

Player in Game

The 'MovePlanar' class is attached to the 'Player' prefab, which is an object that can be instantiated (created) across the network. There are five public variables to allow for the editor to set the values of them before the game is complied. These include speed factor which are used in setting the speed of movement of the player, a rotational sensitivity, which again is essentially a speed factor and the Skin to use for the GUI. There are also a number of Boolean type variables that are used to enable and disable motion and menus depending on inputs such as the 'escape' key, which is captured in the 'Update()' subroutine along with the input and implementation of the movement algorithm. There is also the 'Position', which is of Vector3 type (3D vector), is used for maintaining the position of the player across the network on the clients games. The 'Awake()' subroutine is used to get the RigidBody component of the 'Player' prefab to use for enabling and disabling gravity. The 'OnGUI()' is again used for implementing a menu that is only displayed after the 'escape' key has been pressed with functionality such as exiting the game or enabling motion in the Y axis plane, whilst disabling the gravity option of the RigidBody component. Two other Unity built in subroutines are also used in this class. 'OnSerializeNetworkView()' is used for sending information about the state of gravity and position of the 'Player' prefab across the network to the other players (clients). The second subroutine 'OnNetworkInstantiate()' is used to ensure that the camera is correctly configured to be the one that is a child of the 'Player' prefab.

The 'MoveView' class, shown in the class hierarchy is attached to the camera object which is a child of the 'Player' prefab. The first three public variables are used to set the parameters for the rotational movement in the XY plane. The private 'RotationY' is used to hold the previous value of rotation to be used to rotate the camera. This is not sent across the network as the camera object is not seen by the clients and hence does not require moving on the client side. The last four variables are used to disable and enable the motion of the camera depending on which menus are active. All of this functionality is implemented in the 'Update()' subroutine.

The 'Spawn' class is attached to 'Spawn' game object. This class initiates the 'spawning' (creation) of all of the 'Player' prefabs (one for each connected client and one for the server). The public variable 'PlayerPrefab' is used to store a reference to the 'Player' prefab, which is made using the Unity editor hence the variable being public not private. The 'SpawnPos' is public due to the 'Lobby' class accessing it to give it the player number so that the spawn script spawns all of the players away from one another using the player number to give each a unique spawn position. The 'Awake()' subroutine is used to ensure that the game is connected to/is a server and then initiates the 'SpawnPlayer()' subroutine, which is not a Unity built in subroutine. The 'SpawnPlayer()' then instantiates the current players 'Player' object for them to use as a protagonist in the game. The 'OnPlayerDisconnected()' and 'OnDisconnectedFromServer()' are Unity built in subroutines and are used to make sure the various players are destroyed correctly if a client leaves or if the server disconnects the whole game.

Design in Game

The 'SelectObject' class is attached to the 'Camera' game object like the 'MoveView' class. The top two Boolean types are used to tell all of the subroutines within this class whether an object is within range and being looked at by the camera or whether an object within range of the camera has been selected to move it around. 'Menu' and 'MenuDelay' are used as before to indicate if the menu is being displayed and to stop the 'escape' key initiating the menu appearing an disappearing if the 'escape' key is held down. 'Capture', 'RotationDelay', 'Rotation' and 'CurrentObject' are all used in the movement of an object that has been selected by the player, which is implemented in the 'Update()'

subroutine. 'ObjectMenu', 'ObjectMenuDelay', 'ScrollViewVector', 'InnerText', 'Height' and 'SelectedCategory' are all used to implement the scroll view component implemented as another menu that is used to create objects selected from the menu. The objects contained within this menu are implemented using the public variables 'Categories', 'NumberInCategories', 'ItemThumbnails', 'ItemNames' and 'ItemPrefabs'. These allow for the specification of categories using the 'Categories' variable along with the number of items contained in that category using 'NumberInCategories'. The items three details, their names, thumbnail images and references to their prefabs, are then placed in the same order as the categories in the other three variables. All of this scroll view is implemented in the 'OnGUI()' function up until an item is selected by pressing a button containing the thumbnail of an object. This then runs the 'SpawnFurniture()' subroutine, which creates an instance of the selected prefab in front of the player. This in turn runs the 'SetPositons()' RPC, which moves the object on the other players games to its start position. The 'MoveObject()' RPC is supposed to allow for movement of the objects by sending the name of the object to be moved and also the name of parent (player) that will be moving the object. It then utilises these names to find the objects and tries to move the object with the parent object (player).

Data Communication between BIM Environment and Client End

The realization of BIM data extraction for computer game engine is through using C# based Revit API development. Figure 5 shows the class for BIM data extraction and the extraction process. The 'StartGameEngineViewer' class is used to deal with all of the functionality contained in the plug-in. It is initialised by the button on the ribbon bar created in the 'AddInPanel' class. The first function that is called in the 'StartGameEngineViewer' class is the built in Revit API function 'Execute()'. Therefore, the 'Execute()' function is public because the 'AddInPanel' class needs to access it. The global variables in this class are used mainly due to there being a background worker that runs on a separate thread, which still needs access to some of the data. The 'OBJFile' variable is used to hold a full copy of the '.obj' file for the current building model. The 'uidoc' variable is used to hold a handle to the current document in Revit®. This is so that all of the functions and the background worker can access the document. The 'listener' is object that makes up the 'mini' web-server along with the background worker 'bgWorker'. The 'Domain' and 'PORT' variables are used to hold variables for the settings for the listener. The 'CloseListener' is used so that the main thread can make the background worker thread terminate. The 'AppName' variable is used to hold the name of the plug-in as it may be used in the requests to the 'listener' object and will need to be identified in a requested URL and removed.

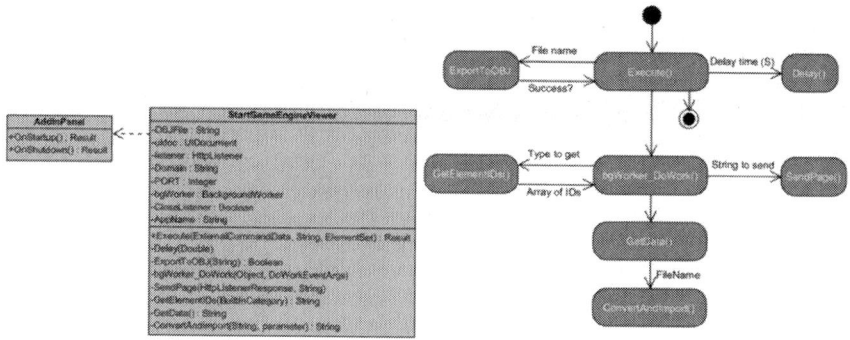

Figure 5: Plugin class and data extraction process.

Figure 6 shows the 'ReadBJFile' and 'OBJModel' classes, which are attached to the 'BuildingModel' game object. The class 'ReadOBJFile' allows for the creation of the building by getting the geometric and some parametric data from the plug-in in Revit® using '.obj' (Wavefront 3D) files. The class 'OBJModel' is used to create an array of the individual objects contained in the building model within the 'ReadOBJFile' class. The 'OBJModel' class essentially is just a container for the variables it needs to hold, which are the ID number and name of the element. It has implemented get and set functions for both of these variables. It also holds an array of 3D vectors (Vector3) for the vertices and the normal's to the faces, which specify the direction in which a face is visible from. 'OBJModel' also holds an array of 2D vectors used to hold the vertex textures or UVs, which are used to position a texture on a face of the model correctly. Finally there is an array of integers that is used to hold a list of indexes to the vertices, normals and UVs used to define each triangle that makes up the mesh of the object.

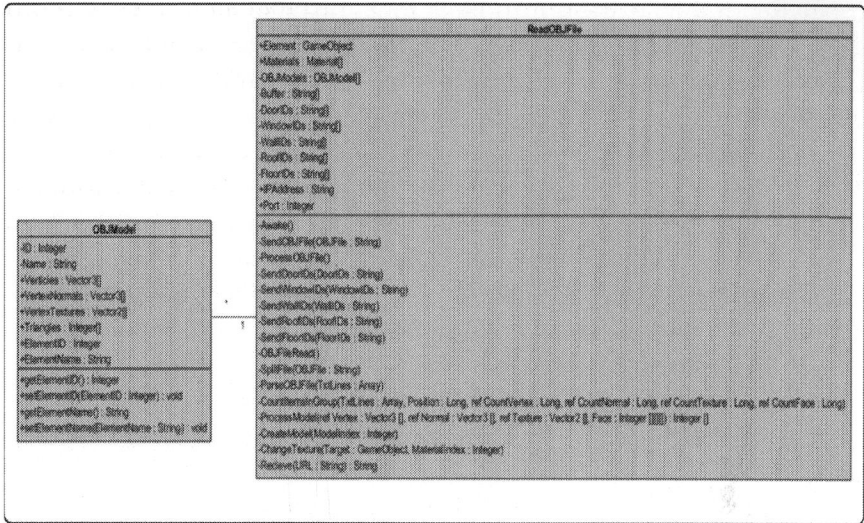

Figure 6. OBJModel and ReadOBJFile classes.

The 'ReadOBJFile' class contains many different variables, functions and subroutines. The first two variables are public so that they can be accessed through the Unity editor. 'Element' is used to hold a reference to the empty prefab only containing the correct components such as a RigidBody to allow for physics interaction this facilitates the building mesh to be built. An instance of 'Element' is used for every element contained in the building. 'Materials' is used to hold an array of type Material that are placed in the array in the editor and are used for applying textures to the various elements of the building. The next private variable 'OBJModels' is an array of OBJModel classes and is used as described to hold the data for the elements ready for creation in the game. Buffer is used to act as a temporary storage location for elements of the '.obj' file that is to be sent to the other players from the server version of the game. The next five private variables are used to hold arrays of strings that will contain element IDs. These are used to apply the correct textures to those five types of elements and to create colliders on only the elements that are not doors so that free movement around the building is possible. The final two variables are public and are used to set where the web client tries to access the building model information ('.obj' file and element IDs).

In 'ReadOBJFile' the only built in function or subroutine used is that of the 'Awake()' subroutine, which is used to initiate other subroutines. The sequence when obtaining a model from the building model 'mini' web-server is as shown in Figure 7. The names of many of the functions and subroutines should be self-explanatory except for the 'CountItemsInGroup()', which is used to count the next set of vertices, vertex normals, UVs and faces of the 3D model, where each set is an element of the building such as a door. Another function that may not be immediately obvious is the 'ProcessModel()' function that is used for transforming the layout of the information in the '.obj' file to the format that Unity uses.

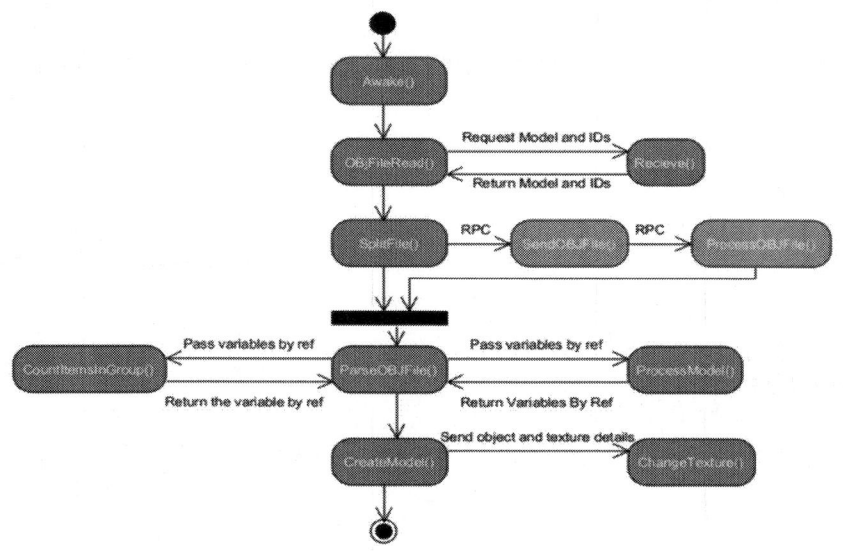

Figure 7: Data transferring process.

RESULTS AND DISCUSSION

System Deployment and Evaluation

In order to fulfil the system development requirements, the following criteria have been concluded for the finished system to be assessed against. These criteria have been evaluated through all stages of the implementation and testing (developer testing) but have mainly been checked using the two test scenarios described later (functionality testing).

The section 4.1 and 4.2 demonstrate how to meet criteria about information communication:

The system must be able to import a building model from Autodesk® Revit® into the Unity 3D based game.

Communication between the Autodesk® Revit® plug-in and the Unity 3D game must be possible.

Properties of elements contained in the Revit® model must be able to be communicated to the Unity 3D game such as parametric properties or element type properties.

The section 4.3 illustrates how to meet functional criteria for end users:

All of the aforementioned items of data (building model, properties, etc.) must be able to be forwarded to clients.

Models of furniture and other items must be able to be placed by an end user on a client version of the game or by the designer into the building model.

Where client (end users) versions of the game place a model in their game the information about that placed object must be sent to the server version of the game.

The designer must be able to transfer information about the objects placed by the clients back to Autodesk® Revit®. The plug-in will then load this information into the building model.

The following sections demonstrate the collaborative design system prototype working as intended, when assessed against the evaluation criteria. All of these tests throughout the testing section were carried out in a networked computing environment running Microsoft® Windows® 7 with Autodesk® Revit® 2012 and Unity 3.4 installed.

Setting Up the Linking between BIM Modelling & Game Engine

The plug-in appears in the ribbon bar of Autodesk® Revit® as a button that will initiate the exporting of the model and set-up the building model server. This plug-in is shown in Figure 8 in the 'Add-Ins' Section of the Revit® Ribbon Bar.

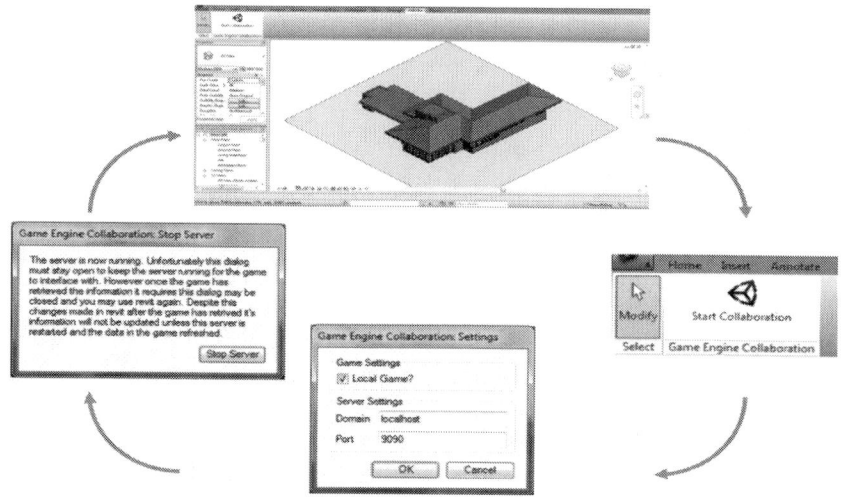

Figure 8: Starting the server to create the link from the game to Revit®.

The form shown in Figure 8 at the bottom, allows for the changing of settings for the 'mini' web-server. These settings include the Internet Protocol (IP) address that the server listens on and also the Port that the server listens to. Finally there is a check box to indicate if the game is to be run on the local machine. This option launches the game in its current location on the local machine automatically after the 'mini' web-server has started. The plug-in also has another form that is displayed, which is shown in Figure 8 on the left of the diagram. This form is displayed due to the nature of the Revit® Application Programming Interface (API), which is a transaction based system. Systems of this nature do not allow the game to retrieve the data required from Revit® whenever it wants to since 'mini' web-server will be started. Once the transaction has completed, Revit® closes the plug-in and the 'mini' web-server with it. To overcome this, the form shown in Figure 8 on the left is displayed to keep the transaction running. This prevents the designer from closing the form but enables the game to obtain its information. Once the game has the information required, the form can be closed along with the 'mini' web-server so that a designer may continue to use Revit®.

Setting Up the Game Server and Game Client

Figure 9 shows the process of setting up a server game and a client game once the plug-in has been started as shown in Figure 8.

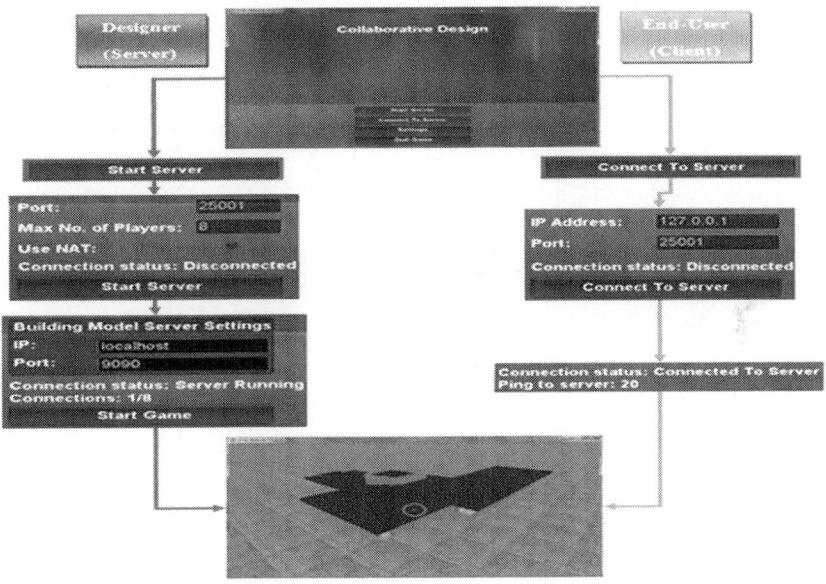

Figure 9: Process to set up a server game and connect a client game.

Testing Scenarios

Testing of the system has been conducted in different ways at different points in the implementation life cycle. Developer testing has been carried out continuously through the implementation of the project. Both component tests and full system tests were undertaken so that the code could be debugged and corrected during the course of the implementation. Once all of the developer testing and implementation has been completed the system was then functionally tested with the two scenarios described below. The first scenario utilised a block of flats created by the designer into which the clients placed objects, which were transferred back to Autodesk® Revit®. This process is illustrated in Figure 10.

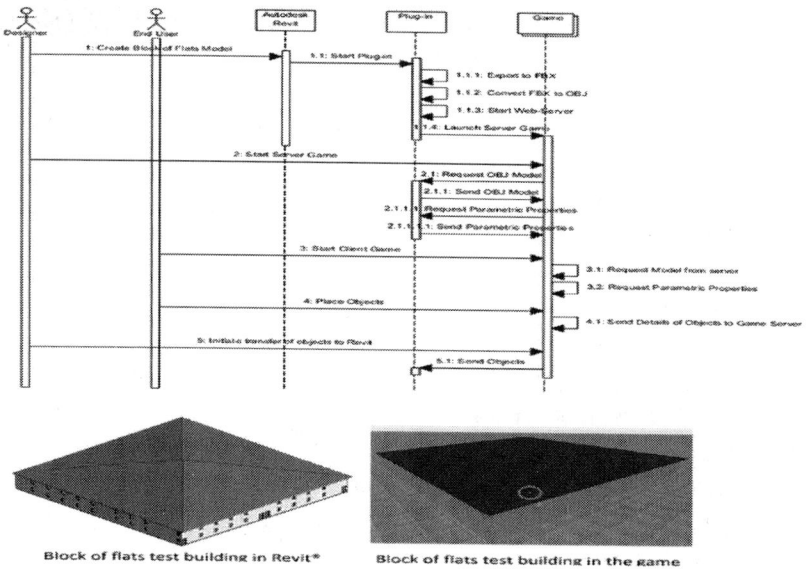

Block of flats test building in Revit*   Block of flats test building in the game

Figure 10: Testing scenario 1.

The second scenario employed a single storey house design created by the designer. Within this house one of the rooms was decided to be a kitchen. The user then placed the kitchen of their choosing into this space. The BIM models and Game environment for this are shown in Figure 11, and the similar sequence modelling (to the one showed in Figure 10) has been used.

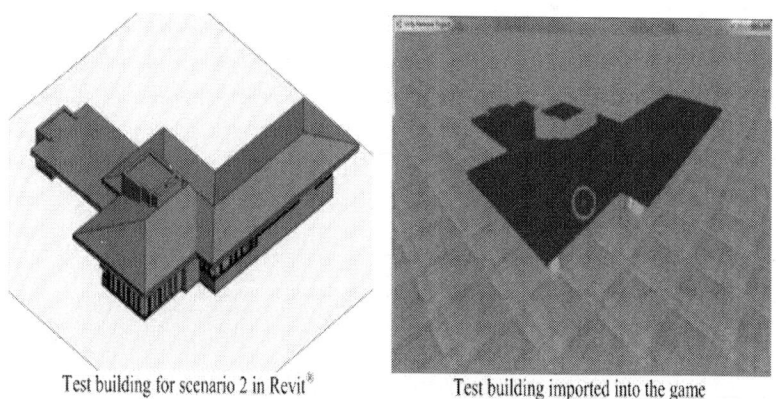

Test building for scenario 2 in Revit®   Test building imported into the game

Figure 11: Testing scenario 2.

The second example shows one new build housing estate including many properties, it allows different clients (end users) allocated to each of the properties to place their respective kitchens. The two previous examples only use the example of furniture as objects to be placed. This is not the only use of the system. It could be used to place machinery in a factory before construction to give a lay out, or to place other plant on a construction site. However, in this paper furniture has been used to demonstrate the system.

Throughout the implementation each element of the plug-in and game has been continuously tested. For example every one of the controls has been checked to see if they performed and completed the expected operations. During the development process, any bugs that were encountered were rectified before the next element was implemented. The main way in which the system as a whole was tested was the use of the two scenarios laid out previously (Figures 9 and 10). These were carried out following the procedure seen in the sequence diagrams previously. For both scenarios it was necessary to create test buildings in Revit. As showed in Figures 10 and 11, it can be seen that the two test buildings in Revit® on the left and then imported into the game on the right.

For the next part of these tests the game was started with a client connected, to place their required objects within the building. The placing of objects in the client, within the block of flats model, can be seen in Figure 12. To achieve this, the server version of the game has successfully imported its own copy of the geometry of the model from the plug-in and then sent this to the client for both the server version and the client version to build up the 3D model within their respective gaming environments.

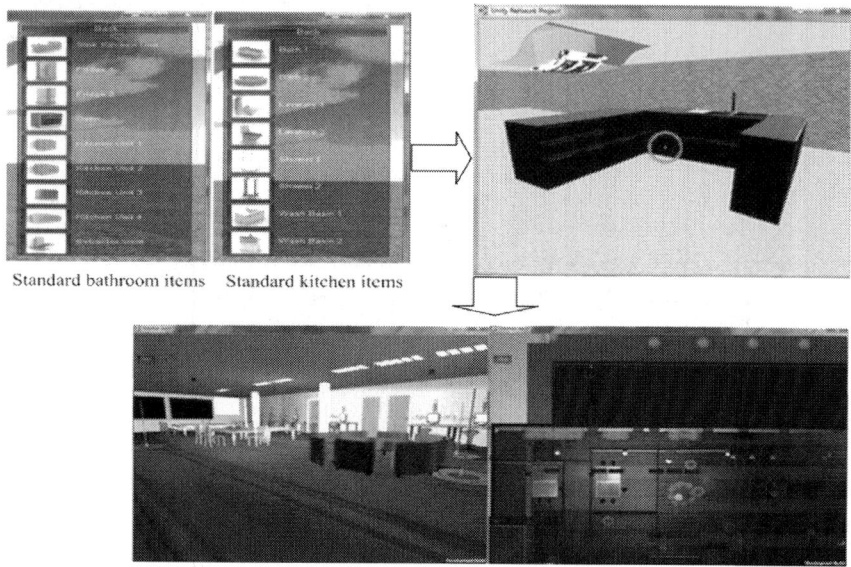

Standard bathroom items   Standard kitchen items

Figure 12: Placing object in the client (right) with the server observing (left) to design a room (bottom).

In Figure 12 it can be seen that the client has placed the objects where he desired on the game screen, these changes (objects positions and rotations) will then be sent back to the server, which will send the changes further back to BIM design environment. This process is initiated on the server game, which the designer would be in control of.

System Evaluations

A total of thirty individuals took part in the test, each using the three different Unity3D clients (including five modes) to finish basic user-centred tasks and completed a pre-experiment and post-experiment questionnaire. The average response was recorded for analysis except for situations where there was a clear bi-polar response when the responses are illustrated individually.

The participant questionnaire responses expanded on the differences between each interface focusing on manoeuvrability, immersive feeling, realistic feeling, quality of model representation and level of interest/excitement (Figure 13). Clearly the greatest difference between the interfaces lied in manoeuvrability. The motion sensor virtual reality (VR) system (i.e. Kinect with HMD) scored a low 2.67, equating to a median between 'Difficult' and 'Very Difficult' on the response scale. There is then a significant jump to public VR system (i.e. 3D Projector with Razer Joystick) at 4.38, landing at the moderate side of 'Difficult' and then another significant jump to the tablet interface which scored 6.67 on the moderate side of 'Easy'. The two desktop interfaces came in as clear preferences above 'Easy' which was noted as 7 on the response scale; the flight mode beating its first person alternative by 0.42 points. It seems that time is needed for general end-users to adapt to the the BIM-VE clients with ad hoc virtual reality devices that are not present in their usual life.

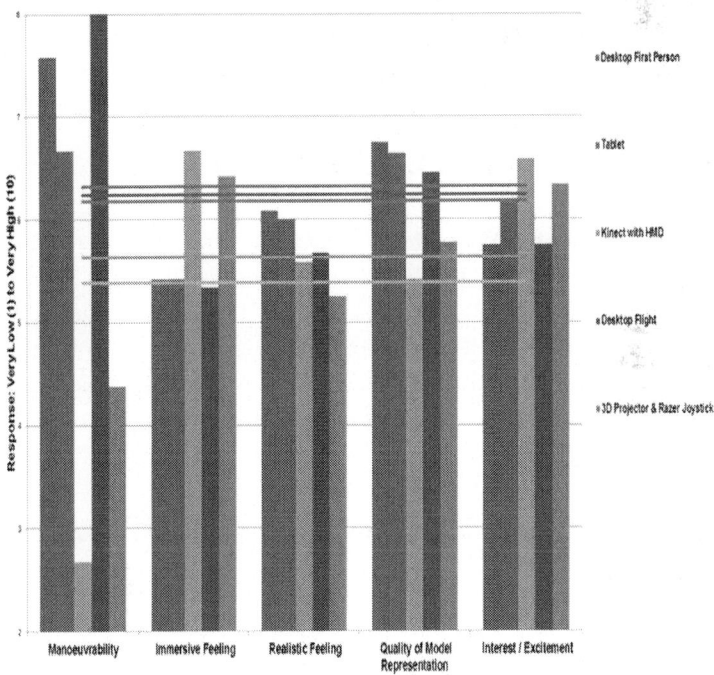

Figure 13: Post-questionnaire response – interface qualities.

There is a slight difference in results for immersive feeling. The tablet and two desktop interfaces showed almost zero differences in immersive feeling whereas there is a noticeable 1.155 average increase with the two VR interfaces, with the Kinect and HMD interface topping the 3D Projector and Razer Joystick system by 0.25 points. It is clear that clients with high quality virtual reality devices can immerse the participants into the virtual environment although the participant requires more time to become familiar with these client devices. In terms of the realistic feeling and quaility of model representation, the participants tended to give the clients they are familar with in their usual life a higher score.

Finally interest/excitement saw both desktop versions achieving the worst score, with the tablet interface scoring noticeably higher by 0.42 points closely led by the 3D projector with Razer Joystick system, all being rated lower than the Kinect & HMD system. This is because that the more accurate details a VE can map with the real via VR devices, the more immersive effects the end users can feel. It demonstates clearly the VR clients of the BIM-VE hold a huge potential to be popular in the future.

CONCLUSIONS

The aim of this paper was to increase end-user involvement in the design process in a collaborative manner. The basics of a framework that could be used to achieve this goal have been designed and constructed into a working prototype that allows a user to interact with a design created in a Building Information Modelling (BIM) program, namely Autodesk® Revit® Architecture. This interaction is achieved by the use of a game engine that aims to provide intuitive controls and an immersive environment to allow non-technically trained persons to engage with the building and add to the design. Unlike other studies that have been conducted into the use of BIM and game engine technology, this prototype enables the automatic import of a project that has been designed in Revit®, directly across a network, into the game with no intermediate design steps. The prototype system has also utilised some of the networking potential of the game engine, in this case Unity created by Unity Technologies, allowing for real-time collaboration between persons in the gaming environment. This has been augmented with the ability for a user to place objects within the model to allow for them to adapt part of the structure with items such as furniture. Objects may be placed successfully in the server version of the game, and similarly objects may be placed in the client version of the game. Worthwhile progress has been made towards what could be an exciting and useful technology framework that the author believes could substantially improve end-users inclusion in the design process.

Regarding future work, the most obvious improvements are to resolve the issues that have occurred with the clients updating the positions of the objects that they have created/placed and relaying this same information back to Revit®. Analysis and debugging to resolve this relaying of information may be relatively straight forward but potentially time consuming task. Another aspect for improvement is to enable the placement of walls and doors to allow for the selection of an internal layout. This could be implemented to work in a similar manner to the way that walls and doors can be placed and used in BIM applications such as Autodesk® Revit®. This would mean that walls and doors had parametric properties and for example a wall would use the properties of the rest of the building such as the floor and ceiling heights to work out where it should start and end in the vertical direction. Likewise a door for example would only be able to be placed on a wall rather than just in space.

Another improvement that may significantly improve the openness of the system and enable substantial improvements in other features would be to modify the system to use IFC files. IFC files are an open format that most BIM applications support and they allow for the transfer of BIM models between different packages. If the system was modified to use IFC files, rather than obtaining the 3D geometric and category/parametric data separately, it would facilitate several improvements. The first improvement would be that more of the parametric data could be extracted from the model permitting uses of the system in other areas such as using it for minor modifications of the structural elements of the building. Using IFCs would also allow for direct updating and integration of new elements into the model, which would simplify the process and potentially speed it up (Bogen and East [2011]).

An additional improvement that would add a great deal of functionality and capability to the system as a whole would be to create a website to host a web version of the game. This would allow users to be authenticated quickly and taken to their specific project in the game, whilst simultaneously improving the ease of access since it would not have to be installed on their computer. It is possible, dependant on how this feature is implemented, that the user would have to install the Unity web player. Alternatively installation of Unity web player could be avoided and the widely available web-based Adobe® Flash® player utilised since version 3.5 of Unity makes provision to create an Adobe® Flash® object. Another web-based improvement would be to implement an interface that permits a user to browse a retailers' website, within the game, to pick out a particular piece of furniture. The game would then download the associated image for the item and seek basic dimensions such as length, width and height, applying this to a generic model of the type of that object.

One further technical improvement could be undertaken would be to implement Universal Plug and Play (UPnP). This permits programs to negotiate with a firewall in a network to allow them access across a network with no user configuration (ISO/IEC 29341–1 [2011]). This would eliminate many of the issues that the system may encounter due to the networked nature of the game and plug-in. This would be largely applicable to the 'mini' web-server and client rather than the game as Unity's networking features already include support for NAT (Network Address Translation) that can bypass many of these problems (Huston [2004]).

AUTHORS' CONTRIBUTIONS

GE - Jointly came up with the idea of creating the BIM-based Virtual environment with LH. Created and programmed the plug-in for Autodesk Revit. Created and Programmed the Unity based 'game' environment. Wrote sections of this paper. LH - Jointly came up with the idea of creating the BIM-based Virtual environment with GE. Wrote sections of this paper. BW - Wrote sections of this paper and reformatted the paper. All authors read and approved the final manuscript.

REFERENCES

1.   AEC (UK) BIM Standard for Autodesk Revit: *Engineering and Construction industry in the UK*. AEC (UK) CAD & BIM Standards Site, United Kingdom; 2010.

2.   Bogen C, East W: *Using IFC Models for User-Directed Visualization in Congress on Computing in Civil Engineering, Proceedings*. American Society of Civil Engineers, Reston; 2011.

3.   BSI: *Constructing the business case - building information modelling*. BSI Corporate, London; 2010.

4.   BuildingSmart: *Investors Report - Building Information Modeling (BIM)*. BSI Corporate, London; 2010.

5.   Christiansson P, Svidt K, Pedersen KB, Dybro U: User participation in the building process. *Journal of Information Technology in Construction* 2011, 16:309-334.

6.   Dickinson J, Woodard P, Canas R, Ahamed S, Lockston D: Game-based trench safety education: development and lessons learned. *Journal of Information Technology in Construction (ITcon)* 2011, 16(2011):119-134.

7.   Eastman C, Teicholz P, Sacks R, Liston K: *BIM Handbook: A Guide to Building Information Modeling for Owners, Managers, Designers, Engineers, and Contractors*. Wiley, Hoboken, NJ; 2008.

8.   El Nimr A, Mohamed Y: Aplication of gaming engines in simulation driven visualization of construction operations. *Journal of Information Technology in Construction (ITcon)* 2011, 16(2011):23-38.

9.   GU N, Nakapan W, Williams A, Figen Gül L: *Evaluating the use of 3D virtual worlds in collabora-tive design learning. In the 13th international conference on Computer Aided Ar-chitectural Design (CAADFutures)*. Icon.Net Pty Ltd, St Leonards Sydney; 2009.

10.   Gu N, Singh V, London K, Ljiljana B, Taylor C: Adopting building information modeling (BIM) as collaboration platform in the design industry. In *The Association for Computer Aided Architectural Design Research in Asia (CAADRIA)*. Icon.Net Pty Ltd, St Leonards Sydney; 2010.

11.   Hassanien Serror M, Inoue J, Adachi Y, Fujino Y: Shared computer-aided structural design model for construction industry (infrastructure). *Comput Aided Des* 2008, 40(7):778-788. Publisher Full Text

12. Huston G: Anatomy: A Look Inside Network Address Translators. *The Internet Protocol Journal* 2004, 7(3):2-32.

13. ISO/IEC 29341–1: *Information technology -- UPnP Device Architecture -- Part 1: UPnP Device Architecture Version 1.0. 2011.* International Organization for Standardization, Geneva; 2011.

14. Lin KY, Son J, Rojas E: A pilot study of a 3D game environment for construction safety education. *Journal of Information Technology in Construction (ITcon)* 2011, 16(2011):69-84.

15. McGraw_Hill_construction: *The Business Value of BIM in Europe.* The McGraw-Hill Companies, Columbus; 2010.

16. Moore ME: *Basics of Game Design.* A K Peters/CRC Press Taylor & Francis Group, Boca Raton; 2011.

17. NBIMS: *United States National Building Information Modeling Standard Version 1 - Part 1 - overview principles and methodologies.* National Institute of Building Science, Washington, DC; 2007.

18. Obituary: Alexander (Sandy) Shafto Douglas 1921–2010 *Comput J* 2011, 54(2):187-188.

19. Petri I, Li H, Rezgui Y, Yang C, Yuce B, Jayan B: *A HPC based cloud model for real time energy optimization, Enterprise Information Systems.* Taylor & Francis, UK; 2014.

20. Revit 2012 API: *Developer's Guide.* Autodesk, Inc, United State; 2011.

21. Rüppel U, Schatz K: Designing a BIM-based serious game for fire safety evacuation simulations. *Adv Eng Inform* 2011, 25(2011):600-611.

22. Shiratuddin MF, Thabet W: Utilizing a 3D game engine to develop a virtual design review system. *Journal of Information Technology in Construction (ITcon)* 2011, 16(2011):39-68.

23. Stuart, K (2011). Modern Warfare 3 smashes records: $775m in sales in five days. Accessed 2 December 2014. [http://www.theguardian.com/technology/2011/nov/18/modern-warfare-2-records-775m] website

24. Sun L, Fukuda T, Resch B: A synchronous distributed cloud-based virtual reality meeting system for architectural and urban design. *Frontiers of Architectural Research* 2014, 3(4):348-357.

25. Wang B, Li H, Rezgui Y, Bradley A, Ong HN: BIM based Virtual Environment for Fire Emergency Evacuation. *The Scientific World Journal* 2014, 2014:589016. doi:10.1155/2014/589016

26. Yan W, Culp C, Graf R: Integrating BIM and gaming for real-time interactive architectural visualization. *Autom Constr* 2011, 20(2011):446-458.

Approaches for Modelling the Energy Flow in Food Chains

Baboo Lesh Gowreesunker and Savvas A Tassou

RCUK National Centre for Sustainable Energy Use in Food Chains, Brunel University, UB8 3PH, Uxbridge, Middlesex, UK

ABSTRACT

Background

The heavy reliance of the global food chain on the use of fossil fuels and anticipated rise in global population threatens future global food security. Due to the complexity of the food and energy systems, the impact of adequate food, climate or energy policies should be carefully examined in a modelling framework which considers the interaction of the food and energy systems. However, due to the different modelling approaches available, it can be very difficult to identify which method best suits the required purpose.

Method

This paper presents the three main modelling approaches as 'top-down', 'bottom-up' and hybrids. It reviews different models under each category in terms of the practicality, benefits and limitations with reference to different past studies.

Results

Bottom-up approaches generally tend to provide high levels of details, but their specificity to particular products/processes detracts their application to holistic models. On the other hand, top-down approaches consider the holistic aspects of the food chain, but the limited level of disaggregation prevents the identification of energy and environmental hot-spots. As a result, hybrid models seek to reduce the limitations of the individual approaches.

Conclusions

This paper shows that the choice of one modelling approach over another depends on a variety of criteria including data requirements, uncertainty, available tools, time and labour intensity. Furthermore, future models and studies have to increasingly consider the inter-dependence of implementing social, demographic, economic and climate considerations in a holistic context to predict both short- and long-term impacts of the food chain.

BACKGROUND

Overall Perspective

The Food and Agriculture Organisation of the United Nations (FAO) has expressed concern over the high dependence of the global food sector on fossil fuels and the projected 70% increase in current food consumption by 2050 due to the rise in global population (FAO [1]). The food sector accounts for 30% of the global energy consumption and 20% of global greenhouse gas (GHG) emissions, with a major contribution from fossil fuels (FAO, [1],[2]). Developed economies such as the UK used approximately 18% of the total energy consumption for the food sector, which produced approximately 32% of the country's GHG emissions in 2011 [3]. The disparity in energy consumption is, however, significant between developed and developing countries, whereby the former use the majority of energy in processing and distribution, whilst the latter use energy mainly for retail, preparation and cooking (FAO [1]).

Due to increased use of depleting fossil fuel resources, the energy-food-climate nexus (FAO, [1]) has been found to be a crucial and complicated challenge for the planet. Energy, food and climate change are intricately linked such that actions taken or policies imposed in one area are very likely to have consequences in the other areas. The nexus can be summarised as follows: high usage of fossil fuels impacts the climate due to GHG emissions - the food sector is heavily dependent on fossil fuels and becoming even more so due to rise in global population - but fossil fuel reserves are depleting and climate change is expected to lower average agricultural yields (Lobell et al. [4]; Roberts et al. [5]). The scenario is therefore complex, and the Food and Agriculture Organization (FAO [1]), the United Nations Environment Programme (UNEP [6]), the Clinton Global Initiative (CGI [7]), Grace Communications Foundation [8] and Mistry and Misselbrook [9] suggest that a reduced dependency on fossil fuels and an increased use of renewable energy technologies are imperative in an attempt to tackle this nexus.

Current barriers to adopting renewable energy technologies are mainly the high capital costs [10] and relatively low-energy efficiencies of some common systems (e.g. ≈3 to 20% for photovoltaic systems [11]). Although various governmental incentives are available, subsidies for renewable energy technologies worldwide are low, hindering their rapid adoption relative to fossil fuels [12]. Nonetheless, renewable energy systems are proving to be cost effective in countries where there is a heavy reliance on diesel electricity-generators and which have high fuel costs (WFP [13]). A study by Maggio and Cacciola [14] conveyed that although there is considerable uncertainty in the lifetime of fossil fuel reserves, the global production of oil will start declining by 2015, whilst gas and coal production will peak in 2035 and 2052, respectively. Historically, the prices of oil and food commodities have been interlinked (FAO [1],[15]; Heinberg and Bomford [16]), but some studies also suggest that in the long run, the price of agricultural commodities have a higher impact on food prices than oil commodities ([17]; Lambert and Miljkovic [18]). This ambiguity in projections can be associated with the assumptions made and modelling approach used in the respective studies [19]. It is generally agreed, however, that due to the current intensive use of fossil fuels in the food sector, the uncertainties in fossil fuel energy availability and prices may threaten food security and affect political stability in the future (FAO [1],[2]). It is therefore imperative that the impact of increased use of renewable energy technologies and the adoption of new more efficient technologies and energy supply systems on food security and supply chain sustainability be investigated further (FAO [1],[20],[21]).

Scope of this Review

This review mainly targets readers interested in furthering their understanding with respect to developing models which study parameters influencing the energy and GHG emissions flows in product-specific, national and international levels of the food chain. It aims at providing an appreciation of existing models and hence informs the reader of the potential benefits and drawbacks of employing different modelling methodology. This review focuses on the food and energy/GHG chain, where the growing field of sustainable consumption suggested that food, home energy and transportation together form a large share of most consumers' personal impacts [22]. Food represents a unique opportunity for consumers to lower their personal footprints due to the high impact of food, high degree of personal choice and a lack of long-term 'lock-in' effects which limit consumers' day-to-day choices [23]. In this regard, Garnett [24] summarised three perspectives on tackling the food security and sustainability issues as: efficiency oriented; demand restraint and food system transformation, of which efficiency oriented measures have been advocated by governments and food industry decision makers [23],[24]. As such, it is imperative to understand the flow of energy in the food chain in order to identify optimum energy efficient and sustainable pathways.

The food chain refers to the successive collection of the farming, food processing, distribution, packaging, retail, catering, household operations, waste and disposal industries, for different types of food products. The emphasis of this review is on the direct and indirect energy consumption and the associated GHG emissions of the overall food chain (or specific food product chains), where the definition of food products abides by the EU foodstuff law (Regulation (EC) No. 178/2002) which refers to foodstuff as 'any substance or product whether processed, partially processed or unprocessed, intended to be, or reasonably expected to be ingested by humans'.

As alluded in the 'Overall perspective' section, the linkages between the food and energy sectors are complex and depend on a variety of factors. The energy systems are currently at a crossroad whereby policies need to determine a balance between sustainable development, competitiveness and supply security [20]. As a result, the interactions involved in the food and energy system should be addressed in a quantitative manner and a modelling framework, so as to aid effective policy design [25]. These models can be employed to evaluate energy effective pathways and the implementation of renewable energy technologies to deliver the energy/GHG emissions reduction targets, at present and in the future. However, in order to allow the proper selection of a modelling method, it is important for the user to understand the particularities of the model. The rationale for this paper therefore relates to the need for examining the benefits and drawbacks of current modelling approaches employed in relation to the food chain and its products and to understand the degree to which such approaches capture the complex and nonlinear interactive behaviour of the food-energy components [25],[26]. The modelling approaches considered in this review are 'bottom-up', 'top-down' and hybrid, and their appreciation will aid in identifying and consolidating new developments in food-energy/GHG models and in elaborating on the performances and discrepancies of current models. In this regard, the paper is divided into the different modelling methods in the 'Methods' section - which describes the models, as well as various studies that employed these models; the 'Results and discussion' section then compares the models and presents and analyses the benefits and limitations of each approach and the 'Conclusion' section then concludes on the general progression of adopted modelling approaches in the literature and suggests future modelling pathways.

METHODS

Bottom-up Approach

Bottom-up approaches adopt a view of assembling the local disaggregated influences in order to determine the global impacts associated with a particular product, process, service or industry. It is a detailed approach and therefore requires compiling inventories of energy, environmental, economic and material inputs for the various processes. The following sections do not attempt at describing the technicalities of the modelling methods, but rather to depict the applicability, practicality and benefits/limitations of these methods.

Life cycle assessment (LCA) LCA refers to a product- or process-based analysis of the GHG emissions and energy consumption, usually employing a 'cradle-to-grave' approach. In the food chain, it considers all stages from the farming/agricultural process through to consumption and waste disposal [27]. This method has been deemed important to examine the intricacies attached to food products or systems, where a current dearth of data exists and where future research is crucial [24],[28],[29]. As LCA and life cycle inventory (LCI) studies are extensive and apply to particular regions, this section will explore studies related to the food chain in the UK. AEA group and partners report on the comparative LCA of seven food commodities procured for UK consumption through different supply chains (DEFRA report no. FO 0103 [30]). The analysis was performed with reference to primary energy use and global warming potential, with assumptions relating to agricultural yields, transportation, and uncertainties due to imports increasing the overall uncertainty of the study. Fisher et al. [31] adopted a cradle-to-retail approach to study the GHG and secondary energy impacts of UK groceries, with special emphasis on high sales-volume products. The food products include alcoholic drinks, ambient products (breakfast cereals, canned food etc.), bakery, dairy, fruit and vegetables, meat, fish, poultry, eggs, non-alcoholic drinks and chilled and frozen products. Data were obtained from various sources including journals, industries, government reports and eco-labels. The study therefore assumes that the different sourced data can be combined together to form a whole chain analysis, but acknowledges the implications of this assumption in the final results of their study. Similar to the previous study, caution is suggested by the authors before using the quantitative results from this study due to the high level of uncertainty. Lillywhite *et al.* [32] studied the embedded energy associated with producing, processing and distributing a range of food products, with a view to addressing the food chain security issue in the UK. Data were also obtained from academic and grey literatures, where the lack of LCA data for multi-ingredient products such as pasta sauce, soup and pizza were derived by the authors' own LCA. This study further explored the price volatility and elasticity of food products with regard to food security and showed that the UK's food supply is almost completely dependent on fossil energy, raising concerns for future food security. The common trait of LCA studies to assume different sourced data can be combined, increases uncertainty in the analysis, where the authors generally caution the reader of the implications in the final results of their study.

In the UK, publicly available specification (PAS) 2050 was developed in 2008 to provide a consistent method for quantifying carbon footprints. The PAS 2050 standard was adopted from the ISO 14044 standard and has been refined with consultation of various research and user communities (PAS 2050 [33],[34]). The stepwise procedures are as follows: (i) defining the 'system boundary' of the product life cycles, (ii) data collection, (iii) compilation and validation of emissions flows and (iv) identification of hot-spots and emissions reduction opportunities. The resource and energy use data are therefore crucial for this method, and primary data generally improve the accuracy (PAS 2050 [34]). This method requires that assumptions - generally relating to primary energy conversion factors, transportation energies, refrigerant leakages, waste disposal and agricultural emissions - are made clear and conservatively. The PAS 2050 method is valid if the assumptions are < 5% of the total footprint and the sample size is adequate (PAS 2050[34]).

Holmes *et al.* [35] conducted a PAS 2050 life-cycle study for five food products in the UK, starting with agricultural production, up to the delivery of food to the catering site, employing past studies' data to evaluate the GHG emissions resulting from raw materials, energy use in manufacture, distribution, retail and wastes. The main assumptions related to GHG emission factors from fertilisers and pesticides, and transportation distances. The study provided emission reduction practices and showed that the level of assumptions, time and resources required pose potential barriers to LCA methods. Tassou *et al.*[36] employed the PAS 2050 method to explore the GHG impacts of food retailing. The report focused on emissions from energy consumption, refrigerant leakage and waste for products ranging from fresh meat to bread. The authors demonstrated that PAS 2050 can be used to quantify GHG emissions from food retail operations if stores are submetered to a sufficient level. They showed that appropriate functional units and boundaries must be selected when studying services with PAS 2050. Campden and Chorleywood Food Research Associates and Partners (DEFRA FO0409 [37]) employed PAS 2050 to study the GHG emissions from the preparation and consumption of complex meals such as cottage pies, bread and apple juice The authors made assumptions relating to the energy use in cooking processes, storage temperatures and amount of water used. This study similarly conveyed that product-based studies should be preferred for PAS 2050, to minimise the uncertainties associated with technology models in process-based studies.

Although LCA methodologies can be applied to products, processes or even industry, it is observed that LCA approaches in the food chain relate mainly to the supply chain of specific food products. Its use tends to focus on GHG emissions rather than energy consumption, which is also an integral part of the assessment. The high level of details possible from LCA can indicate resource efficiency hot-spots and allow targeted actions for improvements [29]. However, the literature suggests that although there are vast LCI of various products, because research is done on a random basis, the compilation of data and hence the comparison of different food supply chains becomes complex (The Ecoinvent database is an example of consistency in LCIs (Ecoinvent [38])). Thus, the system boundaries, assumptions made and uncertainties of the different studies should be clearly specified to increase the potential and practicality of this approach.

MARKAL Model

The MARKAL model is a multi-period linear programming and optimisation tool which allows the simultaneous assessment of several technologies for specific industries or the whole economy (Rath-Nagel and Stocks [20],[39]). The competition between technologies is affected by energy/GHG emission policies, associated costs and technical constraints. MARKAL models can determine the trade-offs between various objective functions such as costs, environmental indicators, oil security, renewable primary energy etc. (Rath-Nagel and Stocks [39]).

MARKAL models are demand-driven models (Rath-Nagel and Stocks [39]), with exogenous energy demands specified by the modeller for a specific network of processes linked through their inputs and outputs, technical, economic and policy based parameters [20]; see Figure 1. As such, the model operates from energy technology databases which detail the current energy system as well as the technical and cost parameters of potential energy systems (Rath-Nagel and Stocks [39]). The model aims to find a partial equilibrium on the energy market which satisfies the maximisation of the net surplus of consumers and suppliers, via linear programming [40]. It should be noted that since the development of MARKAL models by the International Energy Agency in 1976, the MARKAL family has grown to include various improvements. The specifics of the extrapolated models can be found in Loulou *et al.*[40],[41].

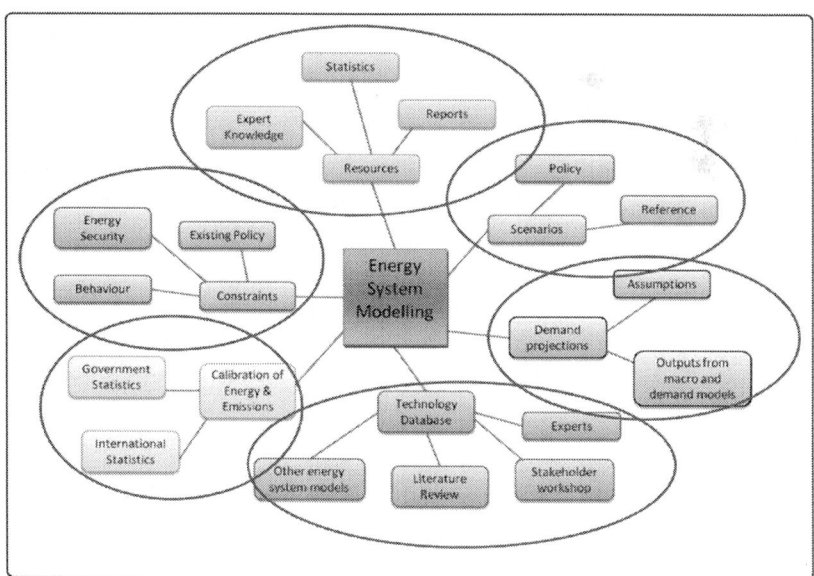

Figure 1: Parameters/database considered in the development of a MARKAL model [42].

The TIMES model is an extension of the original MARKAL model and includes variable time periods, time-data decoupling, higher flexibility energy processes, age-dependent parameters and climate equations (Lolou et al. [41]). Seck et al. [20] employed the TIMES model to analyse the impact of heat pumps on the French food and drink (F&D) industry. The authors used this bottom-up approach as they argue that the French F&D industry ('a non-energy intensive' industry) requires a fine and disaggregated understanding of the existing and emergent technologies. The benefits of adopting such an approach was found to be as follows: (i) the detailed and explicit formulation of the technologies and processes, (ii) the ease of modelling the effect of different policies and (iii) the explicit modelling of the evolution of demand and energy prices. However, the authors state that the TIMES model does not incorporate feedback from other economic sectors and the large quantity of data required for the analysis make the application of the model difficult and perhaps uncertain. Concerning the former point, Loulou et al. [40] argue that the change in energy demand is itself the main economic feedback. Seck et al. [20] assumed an expected evolution of demand for different F&D products, future energy price scenarios and the technical performance of heat pump systems, and showed that the use of heat pumps is a promising technology, possible of 21% reduction in CO_2 emissions and 13.6% reduction in energy consumption in 2020, compared to a scenario without heat pumps.

Gerlagh and Gielen [43] developed a supplementary module 'MATTER 2.0' for the agriculture and food sector of Western European countries. The impetus for this module relates to the increasing importance of competing interest of land use for food, energy and material production, GHG emission reductions and food consumption lifestyle changes. The model assumes a constant mass flow of resources and various proportions of product/energy wastage for different processes, noting that most emissions are converted from the calorific energy content of food. The development of this module shows the versatility of MARKAL for providing a platform for further research and model development.

It should be noted that the MARKAL is now superseded by TIMES, which is supported by the International Energy Agency, and as mentioned before, TIMES has a similar underlying basis as MARKAL, but with several improvements in the techno-economic aspects of the model. MARKAL/TIMES is seen to depict a more aggregated bottom-up approach compared to LCA, therefore allowing a broader overview of the supply chain of various food products or the food chain. The use of MARKAL/TIMES for the food chain is limited in the literatures; however, its use can be promoted especially as it provides a valid and IEA-supported linear optimisation platform for various user-defined scenarios [44]. A more in-depth evaluation is given in the 'Results and discussion' section.

Regression Models

Regression models aim at understanding the causal relationship between two or more variables - where simply adopting correlations may not actually represent this causality - and to quantify how close and well determined the relationship is. It is therefore important in regression models to determine the influential dependent and independent variables, through statistical methods or empirical observations [45].

Spyrou et al.[46] studied the electricity and gas demand drivers for a UK food-retail organisation's buildings using a linear regression method. The authors identified a list of variables that affect the electricity and gas demands based on theoretical concepts and preliminary correlation tests, and observed that the level of detail is limited based on the availability of information. The dependent variables were electricity and gas intensity consumption, whilst the independent variables were divided into physical, operational and regional parameters. The main assumptions relate to the operation and efficiency of the electrical and gas systems, obtained from academic and grey literatures. The authors were able to develop three regression models for the electricity consumption ($R^2 = 0.75$, $p < 0.001$), gas consumption without CHP ($R^2 = 0.62$, $p < 0.001$), and gas consumption with CHP ($R^2 = 0.77$, $p < 0.001$). Employing such a simple model allows the food organisation to quickly identify retail buildings that are under-performing, but does not direct the modeller towards the factors causing the inefficiency.

Boyd [47] developed a performance-based energy efficiency indicator (EPI) for the US food processing sector using linear regression models. Due to lack of data and the diversity involved in food processing, the author adopted a more segmented approach using primary data for specific sectors in the food processing industry, as opposed to modelling the entire processing industry. The study is based on primary energy sources and calculates an EPI for each sector by comparing the actual energy consumption of each processing plant. The optimum scenario is obtained following a stochastic frontier regression analysis, which applies the ordinary least-square regression method to the standard linear regression model. The model predicted R^2 coefficients of 0.8 to 0.96 and variances varying between 0.06 and 0.6. The benefits of using the EPI are that it allows industry leaders to benchmark the performance of various plants and to simply assess the average performance of the sectors, crucial for effective policy designs. However, the large data requirements, the diversity in processing methods and the need for constant updating of the database (due to technical and business innovations) pose limitations in the implementation of such models for longer term.

Tassou et al.[48] employed a regression model to analyse the electrical energy intensity variation with respect to the sales area of UK retail buildings. The study predicted the efficacy of using simple models when the modeller has a firm idea of the influential variables of the system. The authors separated a sample of 2,570 UK food retail buildings based on the sales areas: convenience stores, supermarkets, superstores and hypermarkets. Power law models were employed for each type of store, observing that the electrical energy intensity reduces with increasing sales area, with the relative rate of energy intensity reduction decreasing as the store sales area increases above 2000 m². The standard deviations in the models were found to be higher for small stores (22 to 24%) compared to larger stores (15%). A general relationship was also developed, which showed that the potential for energy savings is higher for small stores, especially if the refrigeration energy consumption is reduced to the mean energy intensity for each category.

Amundson et al.[49] outlined the major steps for developing linear regression models for monitoring and reporting energy savings. The stepwise procedure is illustrated in Figure 2. The authors employed this procedure to develop a model characterising the effects of time resolution on the model coefficients for a large food processing plant and a 'high-tech' factory. The independent variables were the ambient temperatures, production periods etc., whilst total electrical energy consumption was the dependent variable. The authors assessed the fitness of their models using the R^2, coefficient of variation of the root mean square error (CV-RMSE) and model residuals, and quantified the validity of the models using fractional uncertainties. It was observed that the employment of daily resolution improves the uncertainty levels in the models, but the increase in hardware and software requirements and the additional time and complexity of smaller time resolutions should be justifiable. The selection of a suitable model was found to be subject to the opinions of the modeller and the user reaching a compromise.

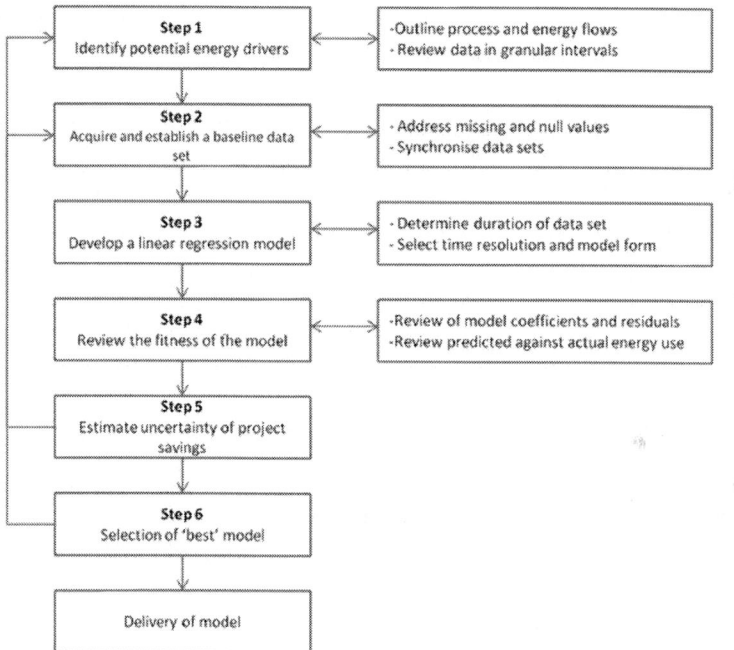

Figure 2: Development process of linear regression model for monitoring and reporting energy savings for industrial maintenance and operation [49].

Regression models are found to represent a simplistic approach to a complex system, by using a set of influential variables. It is important that these models do not only relate these variables but to also include the effects of causality in the representation. The use of regression models (mainly linear regression models) extends in various disciplines (energy, GHG emissions and costs analyses [50]) because of their relative ease of use and assessment of the model errors. The particularities of this approach are evaluated in the 'Results and discussion' section.

'Top-down' Approach

Top-down approaches refer to the decomposition of a scenario into a set of equations where the 'required parameter' is obtained from a combination of the variables considered as 'underlying causes' [51]. The choice of factors depends both on a conceptual model and on data availability. There are two types of equations used in such macroeconometric models: stochastic (or behavioural) and identities. Stochastic equations are estimated from historical data, whilst identities are equations that hold by definition, i.e. they are always true [52]. The variables can be separated into endogenous variables (variables explained in the model) and exogenous (variables imposed on the model) [52], and the model generally applies to an annual basis/time-step. A top-down model can apply to a whole economy or a section of an economy [53]. This section does not attempt to detail the workings of such models, but rather to portray their practicality, benefits and limitations with respect to their use in the food and energy (and/or GHG emissions) chain.

Input-output Model

Economic input-output analysis can be regarded as a collection of the aggregated (intermediate and final) value or amount of goods and services that flows in an economic system and/or as an analytical technique describing and predicting the behaviour of that economic system [54]. The data are usually presented in tabular form and can be obtained from national statistical offices on a yearly basis. Generally, governmental input-output (IO) tables are quantified in terms of monetary value and must therefore be adequately converted energy/emissions values [55],[56]. This is accomplished by following the principle of conservation of embodied energy to create a hybrid monetary-energy table. This principle states that the energy burnt or dissipated by a sector of the economy is passed on, embodied in the product [55]. Since final demand is considered the output of an economy in economic IO analysis, conservation of embodied energy implies that all energy entering an economy is entirely embodied in the final market sales of goods and services [2]. The energetic IO process separates the economy into energy and non-energy sectors and tracks the flow of energy for each sector. It assumes linear homogeneous production technologies, where the output of a sector varies linearly with the production inputs. As such, the concept of energy intensity is often employed [55].

Canning *et al.*[2] employed the IO analysis of the national US food system to trace the energy flow of roughly 400 industries, using data obtained from two federal sources. The study aimed at understanding the factors influencing the US food-energy system over three time periods (1997, 2002, 2007) using a structural decomposition analysis (SDA) and a supply chain analysis (SCA). The SDA evaluated the effects of changes in US population, food-related budget, product mix and food system technologies, whilst the SCA provided a more in-depth analysis of the changes, in terms of the contribution of agriculture, processing, packaging, transportation, retail, food service and household operations to the aggregate change in energy flows. The authors acknowledge that although the boundary of IO analysis is the domestic food system, imported and waste flows are crucial in evaluating the performance of the food chain. Thus, imported embedded energies were obtained by assuming the state of technologies in other countries to be similar US technologies, and waste energy flows were estimated from EPA statistics. The authors observed that energy-intensive technologies accounted for half of the food-related energy increase over the years, whilst the rest was due to increase in population, prepared food and eating out. Household operations accounted for the highest energy use, whilst food processing showed the largest increase from 2002 to 2007, as both households and foods service sectors outsourced manual food preparation and cleanup activities to the manufacturing sector, which relied on energy-intensive technologies.

Zhang *et al.*[57] employed multi-regional input-output model to track the embodied energy for various sectors in China in 2007. A wide range of data were obtained from the Chinese Academy of Science and the National Bureau of Statistics of China, where the linkage between each region and the economy was made by considering the direct primary energy use and imported embodied energies. Amongst the sectors considered, the food production and processing industry accounted for 4.5% of the total Chinese embodied energy use. The study not only showed the potential of IO analyses in terms of flexibility and the number of variables that can be incorporated into the model, but also depicted the statistical knowledge requirements to develop such complex models. Bekhet and Abdullah [58] studied the agricultural energy chain in Malaysia, in an attempt to reduce food imports, minimise energy consumption and increase the yield of the agricultural industry. The authors employed secondary data from National Statistical Databases, considering three energy industries. The study showed a more significant increase in the dependence of the agricultural industry on petrol and coal, compared to the other energy industries, although agriculture is a relatively weak energy consumer. The fisheries sector was found to be the largest consumer of energy, followed by forestry and logging and oil palm estates. The authors acknowledge that because IO data are normally obtained at 5-year intervals, the study fails to identify the changes in energy consumption using a time-series approach, but nonetheless suggest that electricity and gas should be promoted, instead of petrol.

Although the concept of IO analysis was developed for national economic systems, the principles have been extended to specific products. Essengun et al. [59],[60] explored the energy flow of dry apricot and tomato production in Turkey, by collecting primary energy input, quantities and costs of inputs and outputs. The authors employed various multipliers to construct a relatively simple model that allows the modeller to track the energy and monetary flows in the agricultural production. Such model determined the energy efficiency and intensity of the production and suggested different energy improvement measures. Kuswardhani et al. [61] studied the energy and economic IO of greenhouse and open-field production in Indonesia. The authors obtained primary data from surveys and used appropriate conversion factors to obtain the energy values. The study identified the linkages between energy input and crop yield for greenhouse and open-field production, and depicted the energy efficiency ratio of different products. A number of other studies also employed simple adaptation of IO analysis for the agricultural sector [62]-[68]. The primary motivation for choosing the product-specific IO method was the flexibility in collecting primary data, determining energy efficiency ratios and extending the analysis to cost-benefit analyses.

This section has shown that food-energy IO analyses have not only been used at the national or regional level, but also at a more product-specific level. The approaches to IO energy analyses have been to convert national IO tables' monetary values to energy values (using hybrid tables) or to directly employ energy values (or closely related energy parameters). When using the national hybrid monetary-energy IO models, structural decomposition analysis (SDA) has been used to study the sources of changes obtained from IO analysis over different time periods [2],[69]-[72], whilst supply chain analysis (SCA) explores the contributions of different stages of food production to the overall energy flow [2]. The analysis of embodied (i.e. direct and indirect) energy flows has been found to identify and optimise low energy-efficient processes and to propose improvements at both the energy and cost/profitability levels.

Index Decomposition Analysis Models

An index decomposition analysis model refers to the definition of a governing function that relates the aggregate to a pre-defined number of decomposed factors, in order to measure the impact or weight of these factors on the aggregate, over specific time periods. The two most popular approaches for energy analyses are based on the Divisia index and the Laspeyres index where the choice is problem dependent [73]. The Laspeyres index measures changes in an aspect over time by letting the related variables change, but fixing all other variables at the base period values[74]; the Divisia index uses a weighted sum of logarithmic growth rates, where the weights are the components' share in the aggregated value [73].

Hammond and Norman [75] examined the causes and weights of the effects that resulted in the reduction in carbon emissions in the UK manufacturing sector, between 1990 and 2007. They adopted the log mean Divisia index (LMDI) approach, on the basis that LMDI has no residuals and the additive method is easier to interpret. Secondary data were obtained from various sources, and the changes in carbon emissions were decomposed in terms of changes in production volume, inter-sector structure, secondary energy intensity, fuel mix and carbon emission factor. Furthermore, in order to ensure the changes are associated with the correct effects, the manufacturing sector was divided into sub-sectors, which include the manufacture of food and beverages. The individual energy efficiency was taken to be inversely proportional to the energy-intensity effects, which does not imply that improvements in the energy-intensity effects are due to technological effects, unless the energy system is disaggregated to high enough level [76]. The results showed that improvements in energy intensity were the main cause of reduction in carbon emissions in UK manufacturing - whereby more efficient technology, better control and housekeeping, moving towards an increased use of electricity and natural gas, and inter-sector structural change are believed to have contributed to lower energy consumption during the period.

Hasanbeigi *et al.* [77] decomposed the Chinese manufacturing industry using the additive LMDI approach, for both past (1995 to 2010) and future years (2010 to 2020). The authors assumed a constant share of the value-added for each sector, and the future energy projections assumed to be cumulatively decreasing based on the government's reduction targets. The decomposition analysis investigated the effects of aggregate activity, sectoral structure and energy intensity. The additive LMDI approach was selected as it leaves no residuals, and the study showed that the food and beverage sector had the second largest sector rise in value-added from 1995 to 2010, whilst primary energy use remained relatively constant and the energy intensity decreased for the same period. The forecasted energy intensity for the food sector was found to decrease at a lower rate for the period of 2010 to 2020. The decomposition analysis showed that if China intends to meet its goal of re-structuring its economy and moving towards less energy-intensive and polluting sectors, then specific scenarios should be followed. The forecasted decomposition analysis therefore aided evaluation of the impact of the different scenarios on both primary energy use and value-added, at a more disaggregated level.

Generally, the log mean Divisia index model was found to be the most commonly used method[73] for economy, industry or sector level analyses, where the main effects are often separated into production volume, inter-sector structure, energy intensity, fuel mix and/or carbon emission factor. A finer disaggregation level of the dataset allows a more accurate evaluation of the relative impacts of the disaggregated effects, whereby the IEA identified the disaggregated energy indicator hierarchy as shown Figure 3. Furthermore, physical intensities (material, energy and emissions) are suitable, but not preferred for holistic industry decompositions, because of the large varieties in physical units involved. Instead, it is recommended that the energy and GHG intensities be based on monetary units [78].

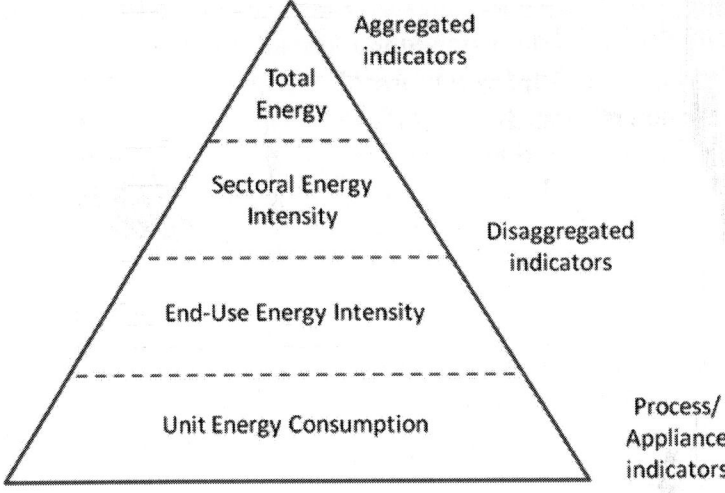

Figure 3: Disaggregated energy indicator hierarchy (adapted from [79]).

In general, the Divisia index model was used in decomposition studies ranging from specific industries, such as the manufacturing industry - which includes food manufacturing [80]-[83], the agricultural industry [84], the service industry [85], to whole economy decompositions [86]-[90]. The Laspeyres index approach was less popular, employed mainly by the International Energy Agency (IEA), due of its ease of use and to ensure consistency in IEA publications (ETO/ESD/LTO, 2007; [79]). The reasons for the lower popularity of the Laspeyres approach were as follows: the model produces residuals when the aggregate values are calculated [74], the choice between the additive or multiplicative Laspeyres model affects the final results of the analysis [91] and the generally lower accuracy of the model compared to the Divisia index method [73].

Dynamic Models

Dynamic models incorporate time into the breakdown of a structural framework using systems of difference or differential equations, in an attempt to provide future expectations/impacts of various policies [92]. These models are more complex and detailed than static models, as they involve assumptions on the rate of economic growth, time preferences, population growth rates, inflation rates, depreciation rates etc. [92],[93]. Researchers to date have adopted dynamic models in different ways and for different applications, but application to the food-energy chain has been limited.

Irz et al. [17] studied the dynamics of price formation of food commodities with respect to agricultural, energy and labour commodity prices, for the case of Finland, by considering attributed relating to the demand and supply sides of the Finnish economy. The authors employed a co-integration analysis using both stochastic and identity sets of equations. The conceptual framework was based on the equilibrium of agricultural, labour and energy commodities for the supply side and disposable income and demographic distribution for the demand side. However, Finland was assumed to have a negligible demographic distribution and disposable income was ignored, due to the unavailability of data. Hence, the empirical model only explained the relationship between food prices and agricultural, energy and labour commodity. The authors studied the system dynamics using a time-series approach by testing for the presence of unit roots and using the Johansen approach to study long-term equilibrium. The causality aspect of the model was analysed using Granger causality tests. The outcome of the model was a long-run relationship of food prices with respect to the other commodity prices, assuming linear growth. The study showed that farm prices represent the main determinant of food prices, followed by wages in food retail and the price of energy. It should however be noted that technological change is implicitly proxied in the time-trend series. Other studies with similar conclusions can be found in Lambert and Miljkovic [18] and Baek and Koo [94].

MAgPIE is a nonlinear recursive dynamic optimisation model, developed by Potsdam Institute for Climate Impact Research Land-use group (PIK [95]). The model requires a set of exogenous demand parameters for each time-step and the yield growth from agriculture is obtained in relation to investment forecasts. The model predicts the impacts of agriculture on land use and GHG emissions. Applications of this model can be found in Dietrich et al. [96], Schmitz et al. [97] or Popp et al. [98].

Hence, dynamic models can be used to study the transient impacts of specific parameters on the economy, but require validation with historical data. The lack of application of dynamic top-down models related to the food-energy chain can be attributed to the fact that the dynamic behaviour and different timeframes (between production and consumption and waste) associated with food is difficult to be forecasted on an aggregated basis due to the large variance in types of food products and technological processes. Furthermore, the level of assumptions that may be required to decompose aggregated energy data for specific food sectors may themselves become a source of question, especially when the temporal aspect is considered. Hence, for such forecasting purposes, hybrid models (such as IO-based LCI or MARKAL-MACRO models) may be more suitable as they allow a higher disaggregation level in terms of technologies and/or food products. Thus, the following section explores the use of hybrid approaches.

Hybrid Approach

A hybrid modelling approach seeks to combine both bottom-up and top-down approaches to allow the modeller to analyse specific details of processes and consider the entire supply chain simultaneously [99]. In the case of the food chain, hybrid IO and LCA were found to be most commonly used in assessing energy and carbon footprints. IO tables provide complete and aggregated data within national boundaries, whilst process-based LCI provide detailed and accurate process information. Suh and Huppes [100] identified three approaches to hybrid IO-LCI methods:

- Tiered hybrid analysis employs process-based LCI data for consumption, waste and upstream activities, whilst the remaining information is obtained from the economic IO-based LCI. Although a simple and fast approach, limitations are (i) the demarcation between process-based and IO-based LCI should be carefully selected, (ii) double counting may occur and (iii) there is no feedback between the two approaches.

- IO-based hybrid analysis requires the disaggregation of the industry sectors in the IO table into sub-sectors and employs the tiered hybrid method for product life cycles outside the IO table boundary. The interactive relationship between pre-consumer stages and the rest of the product life cycles is often difficult to model.

- Integrated hybrid analysis: is a matrix representation of the physical product system, whereby the IO table is connected upstream and downstream of the matrix. The linkages between the product system and the economy can be obtained in terms of purchases and sales records from LCI. This approach puts LCI at the core of the hybrid model and allows full interactions between the individual processes and industries in a consistent framework.

Elgimez et al.[101] employed an IO-based LCI hybrid approach to study the sustainability impact of the US food manufacturing sector. The study consisted of disaggregating the industry into 33 sub-sectors (or product systems), which allowed the consideration of both the direct and indirect effects at a more detailed level than using IO only. In addition to the IO-LCI model, this study also included a data envelopment analysis (DEA) model which evaluated the impact of the sectors through a sustainability performance index (SPI), to allow comparative study. The authors therefore adopted a two-stage hierarchical process where the IO-LCI model provided the environmental outputs, which are then fed to the DEA model to evaluate the SPI of each sub-sector. The authors obtained the imported footprints assuming the same level of technology as the US food manufacturing industry and ignored the regional variations associated with carbon emissions due to the regional homogeneity obtained in IO tables. The study concluded that 19 of the 33 food manufacturing sub-sectors are inefficient and that fisheries and animal slaughtering, rendering and processing are the dominant 'carbon footprint' sectors in the US food manufacturing industry.

Virtanen et al.[102] studied the GHG emissions of the Finnish food-chain using process-based LCA and an IO-based LCA hybrid model. The process-based LCA was obtained from 30 typical lunch plates, used to augment the national IO tables. The hybrid IO-LCA model was derived from a combination of economic input-output tables associated with environmental emissions, LCI of agricultural sub-sectors and related emissions and publicly available LCI and conversion factors for imports (ignoring waste and disposal). Particular emphasis was placed on the agricultural industry because of its relatively high emissions. As a result, this industry was further disaggregated to 44 sub-sectors, with emissions obtained from the ENVIMAT model. The authors observed from the IO-LCI model that the Finnish food chain accounts for 14% of the Finnish GHG emissions, with agriculture accounting for 69% and the energy industry 12%.

Weber and Matthews [23] studied the effects of food miles on the US environmental impact of food. An IO-based hybrid model was employed, where the disaggregation of the transportation sector was done with life cycle information from secondary sources. A commodity-based functional unit of ton-km was employed, where the model assumed that all users of a commodity require the same amount of ton-km per dollar purchase of a commodity, where energy use and carbon emissions were obtained by assuming standard fuel conversion factors. The authors identified limitations of grouping different goods and time lag of data in the IO-LCA model which increase uncertainties in the analysis. Nonetheless, the study depicted that holistic transportation contributes an average of 11% for life cycle GHG emissions and that red meat is more energy-intensive than other food products. The study concluded that dietary shift is a more energy efficient means than 'buying local' in the US.

Wood et al.[103] employed an IO-based LCA hybrid approach to study the comparative energy, water, land and GHG emissions impacts of organic farming and conventional farming in the Australian food chain. The authors conducted a survey of organic farmers to obtain primary data for a process-based LCA for organic farming, whilst conventional farming data were deduced from IO tables. Farm operations were obtained from LCAs and the remaining indirect effects from IO tables. The authors used the SCA decomposition of the hybrid IO-LCA table to obtain the various energy and emissions contributions of the agricultural industry, assuming homogeneous organic farm operations across the industry. The results showed that total embodied energy is generally lower for organic produce, compared to conventional produce, and synthetic chemicals and fertilisers are a major source of energy use, where organic agriculture would reduce these impacts.

In the context of the food and energy/emission chain, the use of hybrid IO tables and LCA models has been found to focus on emissions, rather than energy, because of the relatively higher importance placed by environmental policies on emissions. As such, most hybrid IO-LCA models generally include the details from LCI mainly to the agricultural and manufacturing sections of the food chain, thereby providing more details for these sectors where a majority of the emissions take place in the food chain [101]-[103]. These models have been used as they allow further disaggregation of the conventional economic IO models (as well as including up- and down-stream analyses, beyond the IO model boundary) and incorporate the economy-wide effects to the segregated LCA models. The level of analysis is therefore enhanced, allowing for the detection of environmental hot-spots and better implementation of environmental policies.

RESULTS AND DISCUSSIONS

This paper describes different modelling approaches used in the food chain. These approaches are differentiated as bottom-up, top-down and hybrids, which are qualitatively evaluated as follows, in relation to their applicability to modelling the energy and emissions flow in the food chain.

LCA models provide a process or product based analysis of emissions and energy use in the food-chain using a cradle-to-grave approach. Owing to the complexity of the food chain and the high level of detail possible from LCA, efficiency hot-spots can be identified, therefore targeting actions for improvements [29]. Whilst LCA is good at identifying the intricacies and complex nature of the food chain, this very complexity presents an obstacle to the development of specific recommendations for the future [24]. The level of details required limits the accuracy of this approach and increases the associated uncertainties when such data are not available. Furthermore, it is important to have a standard procedure through which life cycle impacts are measured, with the system boundaries, assumptions, uncertainties and the definition of the functional units of the different studies clearly specified [31],[104]. LCAs need to include ways of measuring outputs that are not only multiple but also intangible - such as social aspects of the chain [24] - and to consider the integration of individualistic models as components to holistic models. Nonetheless, current academic and grey literatures, although performed on a relatively random basis, can provide qualitative assessments of energy and GHG emissions, as well as opportunities for improvements in specific cases [27].

As opposed to LCA models, MARKAL and its sub-models are mainly employed at the national and regional levels, to analyse the implications of different national technology mix ([105]; Teri [106]). The national models and the disaggregated impacts of a particular sector are dependent on the level of detail obtained from the country's MARKAL databases. There are uncertainties when analysing future energy scenarios with regard to the discount rates. The two theoretical concepts of social time preference and social opportunity cost tend to show a divergence in the choice of discount rates between the values adopted by the private sector and the government (Rath-Nagel and Stocks [39]). Hence, different discount rates are often used for different situations and countries (3% for Switzerland, 5 to 10% for the USA, 10% for China or 10% for the UK) [105],[107]. A high discount rate value shows high uncertainty in the energy impacts of a new technology [105].

In most cases, except when the main variables are definite, the development of regression models is an iterative process, requiring a large amount of data and an understanding of the engineering and statistical processes involved. The underlying assumption of linear regression models is that the residuals (difference between predicted and actual data) follow a normal distribution from the mean. This helps to identify outliers (which distort the regression model), model weaknesses and process changes [49]. As such, the judgement of the modeller is crucial in determining the 'best' representative model; e.g. Spyrou et al.[46] considered outliers to be at three standard deviations. Further to the residuals, the fitness and model coefficients are crucial. The use of various indicators such as R^2, variations of root mean square errors, p-value, auto-correlation coefficients, standardised (β) coefficient and fractional uncertainties have been found to be common. Although there are limitations associated with the 'error indicators' [49],[108], the final adoption of a regression model will depend on a compromise between the judgement of the modeller and the satisfaction of the user.

IO models are static linear equilibrium models, where the national IO system boundary for the food-energy chain starts from energy production up to final consumption of food, hence requiring simplifying assumptions to consider the impacts of imports and wastes in the chain. Economic IO tables are usually several years old and in some cases at large time intervals and are mainly used to analyse past energy flows as opposed to predicting these flows. They are therefore not convenient for time-series-type analyses. The results obtained from economic IO analyses are usually exhaustive and applicable to large-scale questions, which are less useful for micro questions [55]. Nonetheless, such an aggregated method usually performed using data from single sources, reduce ambiguities related to acquiring data from different sources. It can also account for the aggregated inter-regional and socio-economic aspects of the chain [109]. When used in the simpler product-specific manner, the efficiency of IO models can be improved when primary data are collected. Furthermore in such studies, the conversion of monetary values to energy values is often not required, eliminating the assumption of sector homogeneity, as energy values are not determined from homogeneous energy prices.

The unique feature of index decomposition analysis (IDA) to provide macro results based on myriad detailed energy indicators gives policymakers quick access to findings from technical data[77]. In IDA analyses, there are issues of data quality, level of sector disaggregation, measurement of output/activity levels and the choice of indicators which would affect the quality and validity of the decomposition results [73] - these are however only dependent on the datasets being used and independent of the actual methodology. Only direct effects can be evaluated with IDA models [110], as opposed to structural decomposition analysis (SDA) that can evaluate embodied effects. An advantage of IDA over SDA is the lower data requirement. However, this is also a disadvantage, since IDA is capable of less detailed decompositions of the economic structure [75],[110].

The comparative aspects of the aforementioned models are presented in Table 1. This comparison has been done in a qualitative manner with respect to the ease of use, benefits, limitations and the assumptions of the models to provide information in the choice of a specific modelling method. Table 1 generally shows that the benefits of one approach are the limitations of the other approach, for instance; bottom-up approaches provide the benefits of high level of detail for a specific product/ sector, whilst top-down approaches have limitations in the disaggregation of the economy to a sector/product level. This therefore led to the development of the commonly used hybrid IO-LCI models.

Table 1: Qualitative evaluation of 'bottom-up' and 'top-down' modelling approaches referred in this paper

Models	Brief description	Common benefits	Common limitations	Common assumptions
Bottom-up approaches				
LCA models	A process or product based evaluation method of the energy use and GHG emissions, typically using a 'cradle to grave' approach	- Measure high quality of energy and GHG emissions data	Level of detail required increases the complexity of data collection	- Secondary to primary energy conversion factors

		- Capture the intricacies and complex nature of the food-energy chain	- Analysis is usually very specific to country and product/process	- Transportation distances and fuels
		- Allow the identification of energy and GHG emission hot-spots	- Information from different sources cannot be combined, unless uncertainties and assumptions are clearly specified	- Imported energies and GHG emissions
			- Choice of functional units and demarcations between system add complexity to the method	- Agricultural sector energies and GHG emissions
			- Ignores the holistic industry impact on the product/process, i.e. static model	- Waste disposal and storage emissions
			- LCA needs constant update	
MARKAL and sub-models	- Demand-driven multi-period linear programming and cost optimisation tools	- Technologies and processes can be explicitly modelled in detail	- Results depend on the accuracy of demand-inputs and description of technological processes	- Description of technological processes
	- Simultaneously assess the impact of several technologies through the partial equilibrium of demand and supply of energy	- Currently being employed in various energy research studies, and various extensions/innovations of the model are being developed	- Discount rates of technologies impact the partial equilibrium when forecasting energy demands	- Discount rates and future energy demands
		- Allows the explicit modelling of the evolution of demand and energy prices	- Does not have information feedback to the wider economy	
			- MARKAL databases have to be regularly updated	
Regression models	Find the causal relationship between the dependent and independent variables in the food-energy chain	- Relatively easy to use and construct	- Require large amount of data to ensure proper correlations	- Determination of errors, and validity of models are subjective to modeller
		- Provide a simplistic description of problem, and allows quick approximations of different policies	- Provide only a quantitative evaluations, and does not show the factors causing the inefficiencies (depends on the level of disaggregation of the model)	- General assumptions relate to the energy consumption and GHG emissions in the data collection phase of the study

			- Assumptions are implicit in the model, hence constant updates required	- Data points outside three standard deviations require further investigation
Top-down approaches				
Economic Input-output (IO) models	Provide the aggregated monetary/energy flow through an economy	- Analyse the impact of the entire economy on each industry, and inter-industry relationships	- Stop at the point of purchase, and ignore waste and imports	- Employ the principle of embodied energy to convert monetary to energetic values
		- Data are usually obtained from same sources, which provides consistency in analyses	- Aggregated data analysis prevents detection of specific energy/ environmental hot-spots	- Economic IO analysis requires energy/GHG assumptions for wastes and imports
		- Quantify the impact/ weight of each sector, and therefore allows identification of low-performing sector	Frequency of national IO tables is low	- Linear production technologies [2]
				- No capacity constraints [111]
				- Sector homogeneity[111]
				Usually assume a constant level of technologies for future analyses [2],[111]
				- Import emissions usually based on domestic production technologies [2]
Index decomposition analysis models (IDA)	Decompose aggregate energy and GHG emissions data into pre-defined factors to measure the relative impacts over specific time periods has been used for: (i) energy demand and supply, (ii) energy-related gas emissions, (iii) material flows and dematerialisation, (iv) national energy efficiency trend monitoring and (v) cross-country comparisons [70]	- Method is relatively quick and simple to implement	- Require an adequate level of disaggregation, else actual effects are not clearly identified	Energy and GHG intensities are usually based on monetary outputs, as opposed to physical outputs
		- LMDI approach has no residuals in decomposition process	- Laspeyres index is simple to implement, but calculates with residuals	
		Provide quick access to assessing the overall impact of policy measures on the economy		
Dynamic models	Aim at predicting future energy and GHG expectations of the food-energy chain	Can provide indication of future energy and policy expectations	- Technological effects are often implicitly accounted in models	- Economic growth rates
			- Can require significant level of assumptions, which questions the validity of such models	- Time preferences

			- Studies related to the food chain are scarce	- Population growth rates
				Inflation and depreciation rates

Gowreesunker and Tassou

Gowreesunker and Tassou *Energy, Sustainability and Society* 2015 5:7, doi:10.1186/s13705-015-0035-y

Such hybrids allow the following: further disaggregation of the economic IO model, incorporate the up- and down-stream analyses and include economy-wide effects to segregated LCA models. The level of analysis is therefore enhanced, allowing for the detection of environmental hot spots and better implementation of environmental policies. Hybrid IO-LCA analyses can pose problems with regard to temporal discrepancies. IO tables are usually published at typically long time intervals (1 to 5 years), making the information in the energy/emission IO tables usually older than a process-based LCA [100] and creating complexities in the clear demarcation of IO and process-based LCA models. From the various literatures reviewed in this paper, it is apparent that the IO-based hybrid approach has been most popular for the food chain, due to the easy availability of national IO tables and extensive amount of LCA studies. However, it is not that one hybrid model outperforms another. The choice of a particular approach is dependent on a variety of criteria such as the following: data requirements, uncertainty of source data, upstream system boundary, technological system boundary, geographical system boundary, available analytical tools, time and labour intensity, simplicity of application, required computational tools and goal and scope of the model [100]. Generally, the use of hybrid IO-LCI models has been found to focus on emissions due to the high importance placed by environmental policies. As such, most hybrid IO-LCA models focus on the agricultural and manufacturing sectors where most emissions take place [101]-[103]. Other hybrid models such as the MARKAL/TIMES-MACRO model from the IEA exist and are worth examining in the context of the food chain.

CONCLUSIONS

This paper presents a review of modelling approaches for the energy and GHG emissions in the food chain. These methods can be classified as follows: bottom-up, top-down and hybrid approaches. The impetus for this study stems from the need to accurately model the holistic food-energy chain, in order to effectively design and implement policies to tackle the food-energy-climate nexus and food security issues. Top-down approaches have been found to consider the impact of the economy on the food-chain and be used to develop national policy measures. However, the limited level of disaggregation due to unavailability of data and the homogenisation of the economy when using top-down models are drawbacks, which do not help in the identification of energy/environmental hot-spots. On the other hand, bottom-up approaches generally provide a high level of detail and capture the intricacies of the food-energy system. However, their specificity to products/processes limits their application to holistic systems if the individual bottom-up models do not follow a standardised procedure. Hence, the predominant trend has been towards hybrid models, which seek to combine the advantages of both bottom-up and top-down approaches. Furthermore, although three hybrid modelling approaches (tiered hybrid, IO-based hybrid, integrated hybrid) have been identified, the IO-based hybrid has been more commonly used in the food-chain.

The choice of one modelling approach over another depends on a variety of criteria including data requirements, uncertainty, technological systems to be modelled, available analytical tools, time and labour intensity amongst others. It should also be noted that the same modelling approach may lead to different results depending on the assumptions made. As a result, the method (i.e. the mathematics and economics), assumptions and limitations of a particular model should be clearly stated in every study. A simple and quick model (such as regression models) may be useful in obtaining a rough indication of policy impacts for specific cases, but when used for the holistic food-energy chain may increase errors and uncertainties due to the complexity of the food chain.

The modelling of the agriculture and waste parts of the food chain was found to involve relatively more assumptions than the other sectors.

This is particularly the case for modelling the GHG emissions because of the need to account for biological processes in agriculture and waste management, based on various conversion factors which increase the probability of inaccuracies in the model. Generally, these inaccuracies can be quantified in various forms, such as R^2 in regression models or relative-percentages in other models.

Most modelling approaches and studies to date consider both the energy and GHG emission aspects of the food-chain. However, in the majority of cases, the emphasis has been on the estimation of GHG emissions and their impact on climate change. Although GHG emissions and energy impacts are complementary in some sections of the food chain, their holistic inter-dependence is not uniform. Hence, the energy part of the nexus is equally important, especially as the food chain becomes more complex, food security becomes more prominent, and food and fossil fuels decouple as more renewable energy systems are implemented in the chain. This study serves as a background to the application of holistic and integrated approaches to modelling the energy and GHG emissions of the food chain. Such approaches have the potential to better represent energy technologies, can integrate different modelling methodologies and can incorporate social, demographic, economic and climate considerations in a holistic context to predict both short- and long-term impacts. Although the application of dynamic models to the food chain was found to be scare, the importance of considering the temporal impact of policy is crucial and requires further research.

AUTHORS' CONTRIBUTIONS

Dr. BLG carried out the review of different modelling techniques presented in this paper, structured and drafted the manuscript. Prof. SAT was involved in finalising the format and content of the manuscript, as well as putting the paper into the overall context of the food chain. Both authors read and approved the final manuscript.

ACKNOWLEDGEMENTS

This study is a result of funding from the Research Councils UK to set up the RCUK National Centre for Sustainable Energy Use in Food Chains

(CSEF), grant no. EP/K011820/1. We would like to acknowledge the contributions of the EPSRC, ESRC and 'Manufacturing the Future' and those of the industry and academic partners in CSEF.

REFERENCES

1.   (2011) Energy-smart food for people and climate. Issue paper, © FAO 20112.

2.   Canning P, Charles A, Huang S, Polenske KR, Waters A (2010) Energy use in the US food system. United States Department of Agriculture Economic research report No. 94.

3.   DEFRA (2013) Food Statistics Pocketbook, 2013, 71 pgs. https://www.gov.uk/government/statistics/food-statistics-pocketbook-2013. Accessed on 12 June 2014

4.   Lobell DB, Schlenker W, Costa-Roberts J (2011) Climate trends and global crop production since 1980. Science 333(6042):616-620

5.   Roberts MJ, Schlenker W, Eyer J (2011) Agronomic weather measures in econometric models of crop yield with implications for climate change. Am J Agric Econ 95(2):236-436United Nations Environment Programme (UNEP) (2011) Towards a green economy: pathways to sustainable development and poverty eradication - a synthesis for policy makers. United Nations Environment Programme www.unep.org/greeneconomy

6.   United Nations Environment Programme (UNEP) (2011) Towards a green economy: pathways to sustainable development and poverty eradication - a synthesis for policy makers. United Nations Environment Programme www.unep.org/greeneconomy

7.   Clinton Global Initiative (CGI) (2012) http://www.clintonglobalinitiative.org/aboutus/tracks.asp, Accessed on 02 January 2014

8.   Grace Communications Foundation (2013) http://gracelinks.org/1309/renewable-energy-at-the-nexus. Accessed on 29 Nov 2013

9.   Mistry P, Misselbrook T (2005) Assessment of methane management and recovery options for livestock manures and slurries. DEFRA report: AC0402.

10. Krozer Y (2013) Cost and benefit of renewable energy yin the European Union. Renew Energy 50:68-73

11. Parida B, Iniyan S, Goic R (2011) A review of photovoltaic technologies. Renew Sust Energ Rev 15(3):1625-1636

12. International Energy Agency - IEA (2012) World Energy Outlook 2012: Renewable energy outlook. http://www. worldenergyoutlook.org/media/weowebsite/2012/WEO2012_ Renewables.pdf. Accessed on 29 Nov 2013

13. World Food Programme (WFP) (2013) http://www.greeningtheblue. org/what-the-un-is-doing/world-food-programme-wfp, Accessed on 29 November 2013

14. Maggio G, Cacciola G (2012) When will oil, natural gas and coal peak? Fuel 98:111-123

15. Kim GR (2010) Analysis of global food market and food-energy price links - based on systems dynamics approach, Hankuk Academy of Foreign Studies, South Korea, Scribd, 18 pages. http://www.systemdynamics.org/conferences/2009/proceed/ papers/P1332.pdf, Accessed on 5 December 2013

16. Heinberg R, Bomford M (2009) The food and farming transition - towards a post-carbon food system, Post Carbon Institute, Sebastopol, California. 39 pages. http://www.postcarbon.org/ publications/food-and-farming-transition/, Accessed on 21 December 2013

17. Irz X, Niemi J, Liu X (2013) Determinants of food price inflation in Finland - the role of energy. Energy Policy 63:656-663

18. Lambert DK, Miljkovic D (2010) The sources of variability in US food prices. J Policy Model 32:210-222

19. Ciaian P, Kancs D'A (2011) Interdependencies in the energy-bioenergy-food price systems: a cointegration analysis. Resour Energy Econ 33(1):326-348

20. Seck GS, Guerassimoff G, Maizi N (2013) Heat recovery with heat pumps in non-energy intensive industry: a detailed bottom-up model analysis in the French food and drink industry. Appl Energy 111:489-504

21. Welsch M, Hermann S, Howells M, Rogner HH, Young C, Ramma I, Bazilian M, Fischer G, Alfstad T, Gielen D, Le Blanc D, Röhrl A, Steduto P, Müller A (2014) Adding value with CLEWS - modelling

the energy system and its interdependencies for Mauritius. Appl Energy 113:1434-1445

22. Tukker A, Jansen B (2006) Environment impacts of products - a detailed review of studies. J Ind Ecol 10(3):159-182

23. Weber CL, Matthews HS (2008) Food-miles and the relative climate impacts of food choices in the United States. Environ Sci Technol 42(10):3508-3513

24. Garnett T (2013) Three perspectives on sustainable food security: efficiency, demand restraint, food system transformation. What role for LCA? Journal of Cleaner Production. In Press, doi:10.1016/j.jclepro.2013.07.045

25. Bazilian M, Rogner H, Howells M, Hermann S, Arent D, Gielen D, Steduto P, Mueller A, Komor P, Tol RSJ, Yumkella KK (2011) Considering the energy, water and food nexus: towards an integrated modelling approach. Energy Policy 39:7896-7906

26. Ingram JSI, Wright HL, Foster L, Aldred T, Barling D, Benton TG, Berryman PM, Bestwick CS, Bows-Larkin A, Brocklehurst TF, Buttriss J, Casey J, Collins H, Crossley DS, Dolan CS, Dowler E, Edwards R, Finney KJ, Fitzpatrick JL, Fowler M, Garrett DA, Godfrey JE, Godley A, Griffiths W, Houlston EJ, Kaiser MJ, Kennard R, Know JW, Kuyk A, Linter BR, et al. (2013) Priority research questions for the UK food system. Food Sec 5:617-636

27. Garnett T (2011) Where are the best opportunities for reducing greenhouse gas emissions in the food system (including the food chain)? Food Policy 36(Sup 1):S23-S32

28. Dron D (2012) Sustainable food, a component of the green economy. In: Corson MS, Werf HMG (eds) Proceedings of the 8th International Conference on Life Cycle Assessment in the Agri-Food Sector (LCA Food 2012), 1–4 October 2012, Saint Malo, France. INRA, Rennes. pp 1-7

29. Mistry P, Cadman J, Miller S, Ogilvie S, Pugh M (2007) Resource use efficiency in food chains: priorities for water, energy and waste opportunities., An AEA Energy and Environment Report to Defra No. AEAT/ENV/R/2457 (ED05226)

30. DEFRA report No. FO 0103 (2008) Comparative Life cycle assessment of food commodities procured for UK consumption through a diversity of supply chains., Report by AEA, Cranfield

University, Ed Moorhouse, Paul Watkiss Associates, AHDBMS, Marintek, submitted to DEFRA

31. Fisher K, James K, Sheane R, Nippress J, Allen S, Cherruault JY, Fishwick M, Lillywhite R, Sarrouy C (2013) An initial assessment of the environmental impact of grocery products. Report submitted to the Product Sustainability Forum (PSF), Code: RPD002-004

32. Lillywhite R, Sarrouy C, Davidson J, May D, Plackett C (2013) Energy dependence and food chain security. DEFRA report FO0415.

33. PAS 2050 (2008) Guide to PAS 2050: how to assess the carbon footprint of goods and services., © Crown 2008 and Carbon Trust 2008, ISBN 978-0-580-64636-2

34. PAS 2050:2011 (2011) The guide to PAS 2050:2011, how to carbon footprint your products, identify hotspots and reduce emissions in your supply chain, © British Standards Institution 2011, ISBN 978-0-580-77432-4

35. Holmes M, Wiltshire J, Wynn S, Lancaster D (2010) PAS 2050 informing Low carbon procurement: pilot study - food, ADAS report to DEFRA.

36. Tassou SA, Hadawey A, Marriott D (2008) Greenhouse gas impacts of food retailing. DEFRA report No. FO405.

37. DEFRA report No. FO 0409 (2008) Understanding the GHG impacts of food preparation and consumption in the home., A report by Campden & Chorleywood Food Research Association, Food Process Innovation Unit (Cardiff Business School), Food Refrigeration & Process Engineering Research Center (University of Bristol)

38. Ecoinvent (2013), The ecoinvent database, Available at http://www.ecoinvent.org/database/, Accessed on the 22 January 2014

39. Rath-Nagel S, Stocks K (1982) Energy modelling for technology assessment: the MARKAL approach. Omega Int J Manag Sci 10(5):493-505

40. Loulou R, Goldstein G, Noble K (2004) Documentation for the MARKAL Family of Models, Energy Technology Systems Analysis Programme. http://www.etsap.org/tools.htm

41. Loulou R, Remne U, Kanudia A, Lehtila A, Goldstein G (2005) Documentation for the TIMES Model PART 1, Energy Technology Systems Analysis Programme. http://www.etsap.org/tools.htm

42. Remme U (2012) Capacity building through energy modelling and systems analysis. Presentation slides from the IEA Experts' Group on R&D Priority-Setting and Evaluation Developments in Energy Education: Reducing Boundaries. Copenhagen 9-10 May 2012. Accessible from http://www.iea.org/media/workshops/2012/egrd/Remme.pdf. Accessed on 01 February 2015

43. Gerlagh T and Gielen DJ (1999) MATTER 2.0: A module characterisation for the agriculture and food sector. ECN-C-99-048, http://www.iea-etsap.org/web/reports/c99048s.html, Accessed on 5 December 2013

44. International Energy Agency - IEA (2011) http://www.iea-etsap.org/web/Times.asp. Accessed on 19 October 2014

45. Ramcharan R (2006) Regressions: why are economists obsessed with them? Finance and Development: a quarterly magazine of the IMF, Vol. 43, No.

46. Spyrou MS, Shanks K, Cook MJ, Pitcher J, Lee R (2014) An empirical study of electricity and gas demand drivers in large food retail buildings of a national organisation. Energy Buil 68:172-182

47. Boyd GA (2011) Development of performance-based industrial energy efficiency indicators for food processing plants. Report to the US Environmental Protection Agency ENERGY STAR Programme, obtained from http://www.energystar.gov/buildings/tools-and-resources/read-about-food-processing-plant-epis, 05 December 2013

48. Tassou SA, Ge Y, Hadawey A, Marriott D (2011) Energy consumption and conservation in food retailing. Appl Therm Eng 31(2–3):147-156

49. Amundson T, Brooks S, Eskil J, Martin S, Mulqueen S (2013). Elements of defensible regression-based energy models for monitoring and reporting energy savings in industrial energy efficiency operation and maintenance projects. In proceedings of the 2013 ACEEE Summer Study on Energy Efficiency in Industry, 25 July 2013, Paper no. 192

50. Esmaeili A, Shokoohi Z (2011) Assessing the effect of oil price on world food prices: application of principal component analysis. Energy Policy 39(2):1022-1025

51. Seibel (2003) Decomposition analysis of carbon dioxide-emission changes in Germany - conceptual framework and empirical results, Federal statistical office of Germany: Environmental economic accounting division. © European Communities, ISBN 92-894-5167-X, 32 pp

52. Fair RC (2004) Estimating how the macroeconomy works. ISBN: 0-674-01546-0, The President and Fellows of Harvard College, Cambridge, Massachusetts, London, England

53. Dagoumas A, Barker T, Scrieciu S, Stretton S (2009) Top-down technological modelling of stabilisation pathways: UKERC scenarios for the UK to 2050 using E3MG-UK. IOP Conference Series. Earth Environ Sci 6:272003 doi: 10.1088/1755-1307/6/7/272003

54. Christ CF (1955) A review of input-output analysis. In Input-output analysis: an appraisal. Princeton University Press (ISBN: 0-870-14173-2), pp. 137–182. Available at www.nber.org/chapters/c2866, Accessed on 21 December 2013

55. Bullard CW, Herendeen RA (1975) The energy cost of goods and services. Energy Policy 3(4):268-278

56. Casler S, Wilbur S (1984) Energy input-output analysis: a simple guide. Resour Energy 6(2):187-201

57. Zhang B, Chen ZM, Zia XH, Xu XY, Chen YB (2013) The impact of domestic trade on China's regional energy uses: a multi-regional input-output modelling. Energy Policy 63:1169-1181

58. Bekhet HA, Abdullah A (2010) Energy use in agriculture sector: input-output analysis. Int Bus Res 3:3

59. Essengun K, Erdal G, Gunduz O, Erdal H (2007) An economic analysis and energy use in stake-tomato production in Tokat province of Turkey. Renew Energy 32(11):1873-1881

60. Essengun K, Gunduz O, Erdal G (2007) Input-output energy analysis in dry apricot production of Turkey. Energy Convers Manag 48(2):592-598

61. Kuswardhani N, Soni P, Shivakoti GP (2013) Comparative energy input-output and financial analyses of greenhouse and open field vegetables production in West Java, Indonesia. Energy 53:83-92

62. Franzluebbers AJ, Francis CA (1995) Energy output: input ratio of maize and sorghum management systems in eastern Nebraska. Agric Ecosyst Environ 53(3):271-278

63. Pahlavan R, Omid M, Akram A (2012) Energy input-output analysis and application of artificial neural networks for predicting greenhouse basil production. Energy 37(1):171-176

64. Ren LT, Liu ZX, Wei TY, Xie GH (2012) Evaluation of energy input and output of sweet sorghum grown as a bioenergy crop on coastal saline-alkali land. Energy 47(1):166-173

65. Ozkan B, Ceylan RF, Kizilay H (2011) Energy inputs and crop yield relationships in greenhouse winter crop tomato production. Renew Energy 36(11):3217-3221

66. Bojacá CR, Casilimas HA, Gil R, Schrevens E (2012) Extending the input-output energy balance methodology in agriculture through cluster analysis. Energy 47(1):465-470

67. Michos MC, Mamolos AP, Menexes GC, Tsatsarelis CA, Tsirakoglou VM, Kalburtji KL (2012) Energy inputs, outputs and greenhouse gas emissions in organic, integrated and conventional peach orchards. Ecol Indic 13(1):22-28

68. Salehi M, Ebrahimi R, Maleki A, Mobtaker HG (2014) An assessment of energy modelling and input costs for greenhouse button mushroom production in Iran. J Clean Prod 64:377-383

69. Rormose P, Olsen T (2005) Structural decomposition analysis of air emissions in Denmark 1980-2002, 15th International Conference on Input-output Techniques Beijing, China June 27 to July 1, 2005.

70. Wachsmann U, Wood R, Lenzen M, Schaeffer R (2009) Structural decomposition of energy use in Brazil from 1970 to 1996. Appl Energy 86(4):578-587

71. Nie H, Kemp R (2013) Why did energy intensity fluctuate during 2000-2009? A combination of index decomposition analysis and structural decomposition analysis. Energy Sustain Dev 17:482-488

72. Cao S, Xie G, Zhen L (2010) Total embodied energy requirements and its decomposition in China's agricultural sector. Ecol Econ 69:1396-1404

73. Ang BW (2004) Decomposition analysis for policymaking in energy: which is the preferred method? Energy Policy 32:1131-1139

74. Cahill CJ, Gallachoir BPO (2012) Combining physical and economic output data to analyse energy and CO2 emissions trends in industry. Energy Policy 49:422-429

75. Hammond GP, Norman JB (2012) Decomposition analysis of energy-related carbon emissions from UK manufacturing. Energy 41:220-227

76. Jenne CA, Cattell RK (1983) Structural change and energy efficiency in industry. Energy Econ 5(2):114-123

77. Hasanbeigi A, Price L, Fino-Chen C, Lu H, Ke J (2013) Retrospective and prospective decomposition analysis of Chinese manufacturing energy use and policy implications. Energy Policy 63:562-574

78. Reddy BS, Ray BK (2010) Decomposition of energy consumption and energy intensity in Indian manufacturing industries. Energy Sustain Dev 14(1):35-47

79. Taylor PG, D'Ortigue OL, Francoeur M, Trudeau N (2010) Final energy use in IEA countries: the role of energy efficiency. Energy Policy 38:6463-6474

80. Jeong K, Kim S (2013) LMDI decomposition analysis of greenhouse gas emissions in the Korean manufacturing sector. Energy Policy 62:1245-1253

81. Hasanbeigi A, Rue du Can S, Sathaye J (2012) Analysis and decomposition of the energy intensity of California industries. Energy Policy 46:234-245

82. Salta M, Polatidis H, Haralambopoulos D (2009) Energy use in the Greek manufacturing sector: a methodological framework based on physical indicators with aggregation and decomposition analysis. Energy 34:90-111

83. Bhattacharya SC, Ussanarassamee A (2005) Changes in energy intensities of Thai industry between 1981 and 2000: a decomposition analysis. Energy Policy 33:995-1002

84. Robaina-Alves M, Moutinho V (2014) Decomposition of energy-related GHG emissions in agriculture over 1995-2008 for European countries. Appl Energy 114:949-957

85. Mairet N, Decellas F (2009) Determinants of energy demand in the French service sector: a decomposition analysis. Energy Policy 37(7):2734-2744

86. Mulder P, de Groot HLF (2013) Dutch sectoral energy intensity developments in international perspective, 1987–2005. Energy Policy 52:501-512

87. Md S, Alam K (2013) Changes in energy efficiency in Australia: a decomposition of aggregate energy intensity using logarithmic mean Divisia approach. Energy Policy 56:341-351

88. Inglesi-Lotz R, Pouris A (2012) Energy efficiency in South Africa: a decomposition exercise. Energy 42(1):113-120

89. Balezentis A, Balezentis T, Streimikiene D (2011) The energy intensity in Lithuania during 1995-2009: A LMDI approach. Energy Policy 39:7322-7334

90. Ediger VS, Huvaz O (2006) Examining the sectoral energy use in Turkish economy (1980-2000) with the help of decomposition analysis. Energy Convers Manag 47(6):732-745

91. Ang BW, Huang HC, Mu AR (2009) Properties and linkages of some index decomposition analysis methods. Energy Policy 37:4624-4632

92. Paltsev S (2004) Moving from static to dynamic general equilibrium economic models (notes for a beginner in MPSGE), MIT joint program on the science and policy of global change, technical note 4 (June 2004).

93. Rout UK, Voβ A, Singh A, Fahl U, Blesl M, Gallachoir BPO (2011) Energy and emissions forecast of China over a long-time horizon. Energy 36:1-11

94. Baek J, Koo WW (2010) Analyzing factors affecting U.S. food price inflation. Can J Agric Econ 58(3):303-320

95. PIK Landuse Group (2011), MAgPIE Mathematical description, obtained from http://www.pik-potsdam.de/research/sustainable-solutions/research/landuse-group/magpie-mathematical-description. Accessed on 14 December 2013

96. Dietrich JP, Schmitz C, Lotze-Campen H, Popp A, Müller C (2014) Forecasting technological change in agriculture—an endogenous implementation in a global land use model. Technol Forecast Soc Chang 81:236-249

97. Schmitz C, Biewald A, Lotze-Campen H, Popp A, Dietrich JP, Bodirsky B, Krause M, Weindl I (2012) Trading more food:

implications for land use, greenhouse gas emissions, and the food system. Glob Environ Chang 22(1):189-209

98. Popp A, Lotze-Campen H, Bodirsky B (2010) Food consumption, diet shifts and associated non-CO2 greenhouse gases from agricultural production. Glob Environ Chang 20(3):451-462

99. Acquaye AA, Wiedmann T, Feng K, Crawford RH, Barrett J, Kuylenstierna J, Duffy AP, Lenny Koh SC, McQueen-Mason S (2011) Identification of "carbon hot-spots" and quantification of GHG intensities in the biodiesel supply chain using hybrid LCA and structural path analysis. Environ Sci Technol 45(6):2471-2478

100. Suh S, Huppes G (2005) Methods for life cycle inventory of a product. J Clean Prod 13(7):687-697

101. Elgimez G, Kucukvar M, Tatari O, Bhutta MKS (2014) Supply chain sustainability assessment of the US food manufacturing sectors: a life cycle based frontier approach. Resour Conserv Recycl 82:8-20

102. Virtanen Y, Kurppa S, Saarinen M, Katajajuuri JM, Usva K, Mäenpää I, Mäkelä J, Grönroos J, Nissinen A (2011) Carbon footprint of food - approaches from national input-output statistics and a LCA of a food portion. J Clean Prod 19(16):1849-1856

103. Wood R, Lenzen M, Dey C, Lundie A (2006) A comparative study of some environmental impacts of conventional and organic farming in Australia. Agric Syst 89(2–3):324-348

104. Garnett T (2006) Fruit and vegetables and UK greenhouse gas emissions: exploring the relationship., Produced as part of the work of the food climate research network, FCRN working paper 06–01 Rev. A.

105. Goldstein G and Tosato GC (2008) Global energy system and common analyses: Final report of Annex X (2005-2008). International Energy Agency (IEA), http://www.etsap.org/FinReport/ETSAP_AnnexX_FinalReport-080915.PDF. Accessed on 5 December 2013

106. (2006) National energy map for India: technology vision 2030. Press, TERI.

107. Strachan N, Anandarajah G, Pye S, Usher W (2010) UK Energy-economic modelling. An overview of the UK MARKAL/TIMES

family., Presentation at the UCL Energy Institute, University College London, July 2010.

108. Reddy TA, Claridge DE (2000) Uncertainty of "measured" energy savings from statistical baseline models. HVAC&R Res 6(1):3-20

109. Martinez SH, Jv E, Cunha MP, Guilhoto JJM, Walter A, Faaij A (2013) Analysis of socio-economic impacts of sustainable sugarcane-ethanol production by means of inter-regional input-output analysis: demonstrated for Northeast Brazil. Renew Sust Energ Rev 28:290-316

110. Hoekstra R, van der Bergh JJCJM (2003) Comparing structural and index decomposition analysis. Energy Economics 25(1):39-64

111. Karkacier O, Goktolga ZG (2005) Input-output analysis of energy use in agriculture. Energy Convers Manag 46:1513-1521

Chapter 4

Intrusion Detection and Big Heterogeneous Data: A Survey

Richard Zuech, Taghi M Khoshgoftaar, and Randall Wald

Florida Atlantic University, 777 Glades Road, Boca Raton, FL, USA

ABSTRACT

Intrusion Detection has been heavily studied in both industry and academia, but cybersecurity analysts still desire much more alert accuracy and overall threat analysis in order to secure their systems within cyberspace. Improvements to Intrusion Detection could be achieved by embracing a more comprehensive approach in monitoring security events from many different heterogeneous sources. Correlating security events from heterogeneous sources can grant a more holistic view and greater situational awareness of cyber threats. One problem with this approach is that currently, even a single event source (e.g., network traffic) can experience Big Data challenges when considered alone. Attempts to use more heterogeneous data sources pose an even greater Big Data challenge. Big Data technologies for

Intrusion Detection can help solve these Big Heterogeneous Data challenges. In this paper, we review the scope of works considering the problem of heterogeneous data and in particular Big Heterogeneous Data. We discuss the specific issues of Data Fusion, Heterogeneous Intrusion Detection Architectures, and Security Information and Event Management (SIEM) systems, as well as presenting areas where more research opportunities exist. Overall, both cyber threat analysis and cyber intelligence could be enhanced by correlating security events across many diverse heterogeneous sources.

INTRODUCTION

Cybersecurity is critical as society becomes increasingly dependent on computerized systems for its finances, industry, medicine, and other important aspects. One of the most important considerations in cybersecurity is Intrusion Detection. In order to mitigate or prevent attacks, awareness of an attack is essential to being able to react and defend against attackers. Cyber Defenses can be further improved by utilizing Security Analytics and Intrusion Detection data to look for hidden attack patterns and trends. Intrusion Detection is also important for forensic purposes in order to identify successful breaches even after they have occurred. For example, it is important to know afterwards if information such as credit card data has already been stolen, in order to take additional precautions or possibly take law enforcement or legal actions. Intrusion Detection can also be very helpful beyond detecting cyber-attacks in noticing abnormal system behavior to detect accidents or undesired conditions. For example, an Intrusion Detection System (IDS) could report anomalies where a malfunction or human error is causing customer credit card numbers to be erroneously charged multiple times. Or perhaps an IDS could alert on something out of the ordinary and detect a gas leak, and help prevent an explosion which could harm or even kill humans. Intrusion Detection can be helpful in providing early warnings and minimizing damage.

This study evaluates some of the advancements in Intrusion Detection technology along with important considerations like monitoring a wide array of heterogeneous security event sources. As cyber-attacks have evolved and grown in sophistication, Intrusion Detection products have also become much more sophisticated, monitoring an ever increasing amount of diverse heterogeneous security event sources. IDSs were the first specialized products developed to detect and alert for potential cyber-attacks, and they can either employ misuse detection or anomaly detection. An IDS utilizing misuse detection evaluates data it is monitoring against a database of known attack signatures to determine attack matches. An IDS utilizing anomaly detection, on the other hand, evaluates data it is monitoring against a normal baseline, and can issue alerts based on abnormal behavior.

One traditional IDS product is a Network Intrusion Detection System (NIDS) which monitors for cyber threats at the network layer by evaluating network traffic. Another traditional IDS product is a Host-based Intrusion Detection System (HIDS) which monitors for cyber threats directly on the computer hosts by monitoring a computer host's system logs, system processes, files, or network interface. An IDS can monitor specific protocols like a web server's Hyper Text Transfer Protocol (HTTP); this type of IDS is called a Protocol-based Intrusion Detection System (PIDS). IDSs can also be specialized to monitor application-specific protocols like an Application Protocol-based Intrusion Detection System (APIDS). An example for this could be an APIDS that monitors a database's Structured Query Language (SQL) protocol. Similar to the heterogeneity of the security event sources such as network and diverse host types, the IDSs themselves can be heterogeneous in their type, how they operate, and in their diverse alert output formats.

Today's Information Technology (IT) security systems and personnel can be inundated with an overload of ambiguous information or false alarms, and the cybersecurity domain frequently encounters problems dealing with Big Data from currently implemented systems. Compounding the problem further, existing IT security systems seldom integrate across a wide spectrum of an organization's information systems. For example, an organization can typically have the following systems: Firewalls, IDSs, computer workstations, Anti-virus software, Databases, end-user Applications, and a variety of other systems. However with traditional IDSs there is rarely any integration among them in the context of monitoring for security breach attempts, and very seldom is there any sort of integrated security monitoring approach across a large proportion of an organization's information systems. A basic illustration of what this paper evaluates is given in Figure 1, where security events from most (if not all) of an organization's computing assets are being monitored. This diagram exhibits the heterogeneity of a typical enterprise's network where security events from different workstations, servers, NIDSs, HIDs, firewall events, etc. can all be very different. For example, an organization might use different NIDS solutions to increase detection accuracy, and increase the heterogeneity of a single function in the security system. To improve Intrusion Detection these security events should be correlated with each other in order to improve alerting accuracy as well as give a more comprehensive overview of cyber threats from an overall perspective.

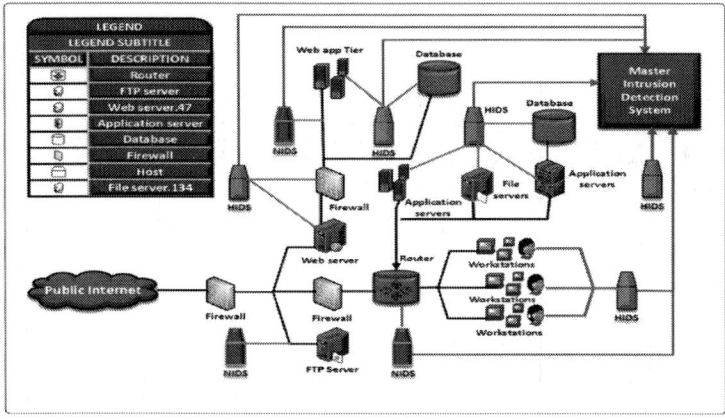

Figure 1: Illustration of Monitoring Heterogeneous Sources.

Intrusion Detection frequently involves analysis of Big Data, which is defined as research problems where mainstream computing technologies cannot handle the quantity of data. Even a single security event source such as network traffic data can cause Big Data challenges. According to Nassar et al. [1], merely 1Gbps of sustained network traffic can cause Big Data challenges for Intrusion Detection while using deep packet inspection. Another Big Data challenge that larger organizations can face is having an incredible amount of host log event data. The Cloud Security Alliance reported [2] that in 2013, it is estimated that an enterprise like HP can "generate 1 trillion events per day or roughly 12 million events per second". They report that such large volumes of data are "overwhelming" and they even struggle to simply store the data. Enterprises dealing with such Big Data issues at this scale cannot use existing analytical techniques effectively, and so false alarms are especially problematic. Additionally, it can be very difficult to correlate events over such large amounts of data, especially when that data can be stored in many different formats. Relational database technology can commonly become a bottleneck in Big Data challenges. For example, commercial SIEMs that use relational database technologies for their storage repositories will find the databases becoming bottlenecks in deployments at larger enterprises: storage and retrieval of data begins to take longer than is acceptable. Zions Bancorporation conducted a case study [3] where it would take their traditional SIEM systems between 20 minutes to an hour to query a month's worth of security data, however when using tools with Hadoop technology it would only take about one minute to achieve the same results. It is a clear sign that Intrusion Detection is facing Big Data challenges when a mainstream technology like relational databases becomes a bottleneck. Next generational Big Data storage technologies like Hadoop can help address these problems.

While traditional Intrusion Detection Systems (IDSs) are a critical component of Intrusion Detection, more focus should be placed on gathering security data from a wider variety of heterogeneous sources and correlating events across them to gain better situational awareness and holistic comprehension of cybersecurity. Analyzing security data across heterogeneous sources can be difficult for Intrusion Detection where homogeneous sources already face Big Data challenges. By analyzing additional heterogeneous sources, the problem can be compounded into a more significant Big Heterogeneous Data challenge as each source can potentially have Big Data. Improving situational awareness by correlating security events or alert data across heterogeneous sources where each can have Big Data challenges is a much more significant problem than performing Intrusion Detection independently on each homogeneous Big Data source, and this is the Big Heterogeneous Data challenge for Intrusion Detection.

A larger IT infrastructure can cause Big Heterogeneous Data challenges with its diversity of input event sources such as various hosts. Correlating among diverse sources like workstations, various application servers, and the network can be a significant problem when facing Big Data challenges. Compounding the problem further is that both the security alerting devices (e.g., IDSs, SIEMS, etc) as well as alert messages can be heterogeneous in nature. The typical enterprise can have a myriad of different security products which do not integrate well, and this heterogeneity causes difficulty for Intrusion Detection. Gartner Research Director Lawrence Pingree addresses this difficulty with a concept called "intelligence awareness" which is the capability of automated intelligence sharing and alerting across a myriad of security systems, and further explains that security systems must become "adaptable based on contextual awareness, situational awareness and controls themselves can inform each other and perform policy enforcement based on degrees or gradients of threat and trust levels" [4]. Ed Billis, CEO of Risk I/O further elaborates on this problem where security products are silo'ed from each other: "SIEMs weren't originally designed to consume much more than syslog or netflow information with a few exceptions around configuration or vulnerability assessment. Security analytics is more than just big data – it's also diverse data. This causes serious technical architectural limitations that aren't easy to overcome with just SIEM" [5].

In addition, industrial processes should also be monitored for Intrusion Detection as industrial systems are increasingly computerized. One example is the nation's electrical grid where most equipment has been computerized that is used to monitor the real-world physical sensors that measure electrical properties like power, voltage, and current. Being computerized, they should be monitored for Intrusion Detection as well. However, the overall Intrusion Detection system can also enhance its capabilities by considering abnormal operational electrical readings and even correlating those real-world events to security events in cyberspace, thus further enhancing situational awareness. Kezunovic et al. [6] discuss the role of Big Data in the electric power industry, and IBM [7] further describes the Big Data challenges faced in this industry along with the need for security monitoring. Clearly, all this Big Data must be monitored in the context of Intrusion Detection. Since real-world physical sensors from the electrical grid can generate Big Data separately, these sensors from the physical world constitute another heterogeneous data source beyond cyberspace and contribute another dimension of heterogeneity as an input to Big Heterogeneous Data. Other industrial applications and processes have increasingly been computerized, with their real-world physical sensors also having Big Data. They can enhance their overall situational awareness by utilizing those physical sensors as inputs into their Intrusion Detection architecture. When doing so, organizations should be aware of the Big Heterogeneous Data challenges they will face.

Even though there have been other survey papers on the Intrusion Detection topic, our paper is unique compared to these prior surveys. We focus on improving intrusion detection from the perspective of aggregating security sensor data from systems and devices which exhibit a great deal of heterogeneity. At the same time, we consider the fundamental Big Data problems that are inherent with such forms of heterogeneous security data. One survey by Modi et al. [8] is especially relevant when considering our work as they focus on Intrusion Detection in the Cloud. Their work does a fantastic job of describing the great deal of heterogeneity of security data and systems encountered in the cloud, and this is increasingly relevant as cloud computing becomes more pervasive and presents more Big Data challenges. Another survey by Zhou et al. [9] is also relevant to ours as they consider heterogeneous architectures for IDSs which collaborate in teams to improve detection accuracy, but they do not consider the Big Data ramifications.

While this paper covers a large variety of issues, there are two main themes of this survey:

- Cybersecurity Data across Heterogeneous Sources.
- Big Heterogeneous Data for Intrusion Detection.

The remainder of this paper is presented as follows: The Intrusion detection and big data background section presents a background on Intrusion Detection and some Big Data implications and challenges. The Security data across heterogeneous sources section covers Security Data across Heterogeneous Sources. The Big heterogeneous data for intrusion detection section discusses Big Heterogeneous Data for Intrusion Detection. The Discussion section provides further discussion and insights about the issues covered. Finally, the Conclusion section concludes the work presented in this paper.

INTRUSION DETECTION AND BIG DATA BACKGROUND

The purpose of this section is to briefly give a general background on Big Data, as well as insight into Big Data challenges facing Intrusion Detection. Some background information is also provided with regards to challenges in security learning such as utilizing publicly available data sets and feature selection. Finally, some examples are provided illustrating how Big Data technologies can be utilized to address Big Data challenges in Intrusion Detection.

Big Data is typically defined in terms of 3Vs, a designation originally developed by Gartner analyst Doug Laney[10] in 2001: Volume, Velocity, and Variety. Volume refers to the amount of data, and there certainly can be a Big Data challenge when large amounts of data pose challenges to processing with traditional computing or techniques (which is also referred to as "Big Volume"). Velocity refers to the speed at which data is processed, and there can be a Big Data challenge when the rate of data is moving too quickly to process with traditional computing or techniques (which is also referred to as "Big Velocity"). Variety refers to the complexity of the data, and there can be a Big Data challenge when the data includes complex problems such as high dimensionality, data from many sources, or data having many different data structures: all of these problems can cause difficulty in processing with traditional computing or techniques (which is also referred to as "Big Variety"). There are many other definitions of Big Data, such as the 5Vs defined by Zikopoulous [11] that adds Veracity and Value to the already existing 3Vs of Volume, Velocity, and Variety. Veracity accounts for the correctness of the data, and can include data quality problems such as noise or missing values (which is also referred to as "Big Veracity"). Value accounts for Big Data in the sense that if particular data does not provide significance (value), it is not relevant for Big Data analysis (which is also referred to as "Big Value"). However for simplicity, Big Data can just be summarized as any time current mainstream computing or techniques cannot process data effectively.

For Intrusion Detection, Big Data is currently a major challenge and has been a prevailing theme for quite some time. In 1994, a study by Frank [12] for Intrusion Detection focusing on data reduction and classification found: "a user typically generates between 3 - 35 Megabytes of data in an eight hour period and it can take several hours to analyze a single hour's worth of data". They further suggested that filtering, clustering, and feature selection on the data is "important if real-time detection is desired," which can improve detection accuracy. This example indicates that Intrusion Detection has been facing Big Data challenges long before the "Big Data" term was introduced.

While a more comprehensive security monitoring system across heterogeneous systems could improve security, it would further exacerbate the Big Data challenge for Intrusion Detection which is already present in isolated systems. Integrating across more security sensors would increase Big Data issues in terms of: Volume in having to store more information collectively, Velocity in that more information would be flowing collectively at a higher rate in and out of the monitoring system, and especially Variety in terms of many different types of information coming from very different sources and also collectively yielding higher dimensionality.

A more comprehensive approach for monitoring a myriad of diverse heterogeneous event sources for Intrusion Detection can yield a better situational awareness of the threats in cyberspace, and thus improve detection accuracy and minimize false alarms by correlating security events among these diverse sources. Experiments have indicated that embracing a more diverse heterogeneous approach to Intrusion Detection does yield better situational awareness and improve accuracy. However, Big Data challenges already exist in some of the individual sources, and when they are aggregated the existing Big Data problem is compounded into a more significant Big Heterogeneous Data problem. When Big Data challenges are already present in any of the underlying inputs or outputs for Intrusion Detection, the overall system will likely experience Big Data challenges as well unless the Big Data bottleneck is eliminated. One way to remove this Big Data challenge is by filtering out (removing) the Big Data from a subsystem. However this is not ideal if valuable information is lost. New techniques or Big Data technologies can alleviate the challenges and costs that Big Data impose for Intrusion Detection.

Big Heterogeneous Data Definitions

When Big Data is present in heterogeneous forms, it can be considered Big Heterogeneous Data regardless of whether that data is input(s) or output(s) of the system. For example, this can arise due to the additive properties of Big Data. If one input is deemed Big Data and is added to another input which is not Big Data, the result will still be Big Data. This can be shown in Equation 1 below:

$$BD(``BigData") + NBD(``NotBigData") = BD(``BigData")$$

(1)

Similarly if some advanced data correlation (or data fusion which is presented in the Security data across heterogeneous sources section) for analysis is occurring and the Big Data is being combined with "Not Big Data" in a multiplicative manner, the result will still be Big Data. This can be shown in Equation 2 below (assuming "Not Big Data" is greater than one):

$$BD(``BigData") \times NBD(``NotBigData") = BD(``BigData")$$

(2)

Therefore, when Big Data is being combined with other data that is not classified as Big Data, the result will still be Big Data.

Another important consideration is that Big Data Challenges can quickly escalate into a significantly larger Big Data problem when combining multiple heterogeneous sources for analysis where each of the sources can have Big Data challenges individually. An example of this would be if two or more heterogeneous sources which separately contain Big Data challenges individually were then analyzed with advanced data correlation techniques (or data fusion which is presented in the Security data across heterogeneous sources section) in order to give better accuracy through superior situational awareness. For complex systems such as Intrusion Detection where a large amount of heterogeneous sources are common and can contain Big Data challenges, the problem can quickly escalate into a more difficult Big Heterogeneous Data challenge. This can be shown in Equation 3 below (where n refers to the number of heterogeneous data sources that contain Big Data, and $n>1$):

$$BHD(``BigHeterogeneousData") = \prod_{i=1}^{n} BHDSi(``BigHeterogenousDataSourcei")$$

(3)

The above generalizations do not always apply and even if parts of the system (e.g., a subsystem) contains Big Data challenges, these do not always propagate throughout the rest of the system. Big Data can be effectively removed in one or more of the subsystems by filtering (removal), and then the Big Data would not necessarily propagate throughout the rest of the system. This is not always an ideal approach if the Big Data being filtered out contains value, but it is still necessary at times if retaining the Big Data is too costly. An example for this would be if netflow traffic was analyzed for a NIDS instead of deep packet inspection. The deep packet inspection will yield superior detection accuracy. However the cost may be prohibitive in doing so. Another example might be the time retention policy for very detailed forensic data, where costs can prevent this Big Data from being stored indefinitely. This is illustrated in Equation 4 below (where the subtraction operator is essentially filtering or removal of the Big Data):

$$BD(\text{``}BigData\text{''}) - BD(\text{``}BigData\text{''}) = NBD(\text{``}NotBigData\text{''})$$

(4)

As the above scenario is a cost and benefit tradeoff, Big Data challenges can also be removed by some of the following "Big Data Handlers": Big Data technologies, natural technology evolution (e.g., Storage or Processing evolutions such as Moore's Law), or novel techniques (or new approaches). For example, if it is desired to retain forensic data longer and a "Big Data Handler" technology like Hadoop permits this to be performed in a cost permissible fashion, then the "Big Data Challenge" can be removed and the "Handled Big Data" can be retained in a manner that is within cost constraints. This is illustrated in Equation 5 below (where the addition operator is enabling the Big Data to be handled in a cost effective manner):

$$BDC(\text{``}BigDataChallenge\text{''}) + BDH(\text{``}BigDataHandler\text{''}) = HBD(\text{``}HandledBigData\text{''})$$

(5)

Essentially, when dealing with Big Data challenges and heterogeneous inputs or outputs, the resulting data will still be Big Data if the Big Data Challenge is not eliminated in some way. For example, if there was a "Big Data Challenge" like a particular data source that had very high dimensionality and if a "Big Data Handler" like feature selection could be effectively used to create "Handled Big Data", then the "Big Data Challenge" will either remain or be eliminated depending on whether an effective "Big Data Handler" was used. In this case, if feature selection was effectively employed as a "Big Data Handler", then the "Big Data Challenge" would be removed and we would have "Handled Big Data". Likewise, if feature selection was not effective (or used), then the "Big Data Challenge" would remain and we were not able to effectively handle the particular challenges from Big Data.

Accordingly, when considering Big Data with heterogeneous sources or outputs, it can be better described as Big Heterogeneous Data so long as the Big Data Challenges are not eliminated in some way. The reason Big Heterogeneous Data is a more descriptive term is because typically the Big Data Challenges will be even more pronounced (i.e., magnified) when dealing with extreme heterogeneity in the input(s) or output(s) of Intrusion Detection Big Data within cyberspace. The Big heterogeneous data for intrusion detection section will further elaborate on Big Heterogeneous Data in terms of inputs and output categories now that the rationale behind the Big Heterogeneous Data terminology has been presented. This Big Heterogeneous Data challenge can become more pronounced while attempting to enhance Intrusion Detection through superior situational awareness by adopting more heterogeneity in the inputs, outputs, and architectural components as mentioned throughout this survey.

Important Considerations for Intrusion Detection

With a more comprehensive security monitoring system, improvements to computer security do not need to be restricted to merely detecting security intrusions. Such a system could be extended to actually prevent security intrusions by integrating with technologies such as Intrusion Prevention Systems (IPSs), and embracing more of a "Defense in Depth" strategy [13]. Naturally, an IPS would require close to real-time detection. Note that this study is not limited to Intrusion Detection with the "real-time" distinction, and also includes offline forensic and security analytic capabilities.

This survey is not similar to previous Intrusion Detection surveys in that it evaluates the Intrusion Detection problem with an emphasis on aggregating security sensor data across many different systems and devices with the motivation of further improving security alerting accuracy. In addition, we consider the Big Data issues that arise when handling such forms of heterogeneous security data.

In the earlier days of computing, security monitoring was primarily performed by system administrators checking the log files of their servers. Then in the 1980s, the concept of the Intrusion Detection System was introduced, where a separate monitoring device would look for suspicious behavior at the network or computer host level. Denning [14] produced what many consider to be the first landmark research paper for Intrusion Detection System (IDS) research back in 1987. A good example of an IDS is the common and widely known open source IDS called Snort [15],[16].

Intrusion Detection is a very active research area with important implications. The Center for Strategic and International Studies and McAfee conducted a study [17] and analyzed monetary losses from cybercrime and cyber espionage: "for the US, for example, our best guess is that losses may reach $100 billion annually." They approximate these global losses to be about $300 billion annually. In 2012 through more than 250 client engagements, the Verizon RISK Team [18] found over 47,000 confirmed security incidents, with 92% of data breaches perpetuated by outsiders in their engagements with clients. In a study [19] by the Ponemon Institute, for the FY 2012 it was determined that the most expensive cybercrime category was "Detection" costing 26% (followed by: Recovery, Ex-post Response, Containment, Investigation, Incident management). These studies clearly demonstrate that cybersecurity (and specifically Intrusion Detection) have significant economic impact.

Julisch and Dacier [20] discuss how Intrusion Detection can have many false alarms: "IDSs can easily trigger thousands of alarms per day, up to 99% of which are false positives." It is not uncommon for security analysts to grow numb to a flood of meaningless false alarms. Xu and Ning [21] state that in terms of detection rate, IDSs are typically not completely accurate, and can have an unacceptable number of False Negatives ("may miss some attacks").

Intrusion Detection is inherently a Big Data problem according to Suthaharan and Panchagnula [22]: "However the biggest challenge is the 'Big-Data' problem associated with the large amount of network traffic data collected dynamically in the intrusion detection dataset". Bhatti et al. [23] discuss how even current technologies cannot cope well with the Big Data challenges of Intrusion Detection: "Security analytics in a big data environment presents a unique set of challenges, not properly addressed by the existing security incident and event monitoring (or SIEM) systems that typically work with a limited set of traditional data sources (firewall, IDS, etc.) in an enterprise network". From a study by Enterprise Strategy Group at the end of 2012, Olsten [24] discusses that: "44% of enterprise organizations consider their security analytics 'big data' today, while another 44% believe that their security analytics requirements will be regarded as 'big data' within the next two years". Clearly, Intrusion Detection can be a Big Data challenge.

Challenges of Machine Learning in Cybersecurity

This section addresses some issues found with Intrusion Detection data set challenges and feature selection. This background is very relevant for Intrusion Detection in general especially considering the widespread criticism of the publicly available data sets, and yet the majority of the research uses these criticized data sets. Accordingly, it is important that the reader understands that many of the experiments considered in this paper suffer the same criticisms when using an experiment with data sets (unless otherwise noted). Some researchers openly admit that the underlying data sets themselves have inherent flaws. Nonetheless, researchers continue to use them because essentially it is the best they have to work with.

A brief background is also given on feature selection and its application to Intrusion Detection data sets. This too can be important for considering experiments with data sets for Intrusion Detection, and especially more so when utilizing data sets from a multitude of diverse heterogeneous sources. The reason for this is the landscape for cybersecurity can change extremely rapidly, and accordingly can undermine the stability of the feature selection sets in general and especially so for real-time Intrusion Detection. These issues will also be evaluated from the perspective of Big Data.

Intrusion Detection Data Set Challenges

Many authors discuss the problems with existing public Intrusion Detection data sets: for example, Sommer and Paxson [25] give an excellent summary regarding some of the underlying reasons why this is a significant problem. The cybersecurity landscape has changed significantly over the last decade, and many don't even consider experiments that use older data sets to even be relevant today according to Sommer and Paxson. Unfortunately, organizations can be reluctant or even legally constrained from divulging sensitive data that these types of data sets can contain, and Coull et al. [26] note that attempts to anonymize sensitive data are not always effective. In addition, lab simulation of real-world network traffic to generate data sets is often not very realistic. As shown in a survey by Azad and Jha [27], the two most popular data sets used for research in Intrusion Detection are DARPA and KDD Cup where out of the 75 studies discussed, 46 used one of these data sets while only 29 chose a different one. This is disappointing because the DARPA and KDD Cup data sets are over a decade old, and even recent studies still frequently use them. The cybersecurity landscape has changed significantly over the last decade, and many don't consider experiments that use those data sets to be relevant today [25]. To make matters even worse, it became widely known shortly after the initial release of those data sets that they contained inherent flaws, as discussed by McHugh [28] and Mahoney and Chan [29]. Nonetheless, the DARPA and KDD data sets are still widely used even today. Some of the more commonly used data sets can be seen in Table 1.

Table 1: Summary of popular datasets in the intrusion detection domain [[30]]

Data source	Dataset name	**Abbreviation**
Network Traffic	DARPA 1998 TCPDump Files	DARPA98
	DARPA 1999 TCPDump Files	DARPA99
	KDD99 Dataset	KDD99
	10% KDD99 Dataset	KDD99-10
	Internet Exploration Shootout Dataset	IES
User behavior	Unix User Dataset	UNIXDS
System call sequences	DARPA 1998 BSM Files	BSM 98
	DARPA 1998 BSM Files	BSM 99
	University of New Mexico Dataset	UNM

Zuech et al.

Zuech et al. Journal of Big Data 2015 2:3, doi:10.1186/s40537-015-0013-4

These flawed public data sets lack Veracity from the perspective of Big Data, and so they would not be relevant as a consequence of having poor quality. Due to this low Veracity, these data sets would also lack Value as well, further reducing their relevancy.

A few other data sets of note are: ISCX [31], MAWI [32], NSA Data

Capture [33], and the Internet Storm Center[34] (which also hosts the dshield.org data set). However, these more recent data sets are not used as frequently as the DARPA and KDD data sets, even in recent studies.

Song et al. took an interesting approach [35] in building their own data set by using honeypot data. A honeypot is a system that is not completely patched in order to draw attention from attackers. They also placed a machine in the network to generate normal traffic, and so any activity related to that machine was labeled as normal (it did not receive much attack traffic) while all traffic related to the honeypots were labeled as an attack. Overall there were approximately 93,000,000 total sessions generated, with about 50,000,000 being normal sessions and the remainder being attack sessions. Also of interest, was that their IDSs and anti-virus did not successfully classify about 426,000 of the sessions as attacks, even though they were able to more correctly classify them as attacks upon deeper inspection of the shellcodes. Because this data did not come from an actual client or pertain to ongoing business efforts, Song et al. did not need to worry about the sensitivity of the data. Also, their data set is roughly balanced between the classes of normal and attack.

While there are some drawbacks to this approach (for example, the normal class could be considered too "simulated"), it shows good promise for future work. More researchers could take this or a similar approach instead of continuing to use datasets that are not very relevant from over a decade ago. Another option for researchers to generate more adequate data sets might be for them to actually launch attacks in honeypot networks with simulated normal traffic, or possibly even do so in a real-world environment if they can properly sanitize sensitive data or ensure the absence of sensitive data in the first place. Generating data sets at even larger scales with honeypots could also lead to Big Data challenges in terms of Volume, Variety, and even Velocity in having to accommodate such large amounts, speed, and variety of Intrusion Detection data.

Intrusion Detection and Feature Selection Opportunities

Feature selection is an important technique in addressing Big Data challenges posed by Intrusion Detection, and when applied properly it can significantly improve classification processing times. In some cases, it can even improve classification accuracy by removing misleading noise. However, one should take caution in how feature selection is applied and especially with regards to research studies versus real-world application in terms of both relevancy and efficiency.

Many studies (even recent studies) are essentially using static data sets (i.e., DARPA, KDD, etc.) in the sense that the labeled instances might not anticipate newer real-world attacks such as "zero-day exploits", and this is especially important in the domain of cybersecurity because it is inherently a dynamically changing landscape. Newer attacks can be significantly more diverse than old attacks in terms of both technical implementation as well as the underlying methods themselves in the ongoing arms race between attackers and defenders. So in terms of feature selection in Intrusion Detection, yesterday's selected features from a static data set might not be relevant for tomorrow's dynamically different data set. A new attack class can make different features important, and different feature sets may or may not be relevant even at the millisecond scale. Thus, it is important to rethink the relevancy of feature sets from data sets that are older, lack diversity, or are very static.

A smaller number of relevant features will improve classification processing times from an efficiency standpoint. However, the process of performing feature selection in some cases can take a considerable amount of computation time. Applying feature selection to a data set where its attributes can change both rapidly and diversely might not be able to generate feature sets in close to real-time. In certain scenarios where generating feature sets takes too long for the effectiveness of the system, perhaps generating feature selection sets could be delegated to an offline process similar to what Bass proposed for offline Data Mining [36]. Those feature selection sets could be applied by the Intrusion Detection templates which are then used at the various sensors. In this manner, some static and stable feature sets could be used for Intrusion Detection. However, it cannot be assumed that all feature selection sets will be stable, especially when they are built from a myriad of heterogeneous sources in a constantly evolving and hostile environment where the diversity in attributes of data sets can vary considerably.

Wang et al. [37] conducted an experiment employing feature selection, specifically to address the so-called "dimensional disaster" problem which often prevents multi-sensor fusion from being applied to Network Security. In their experiment the original number of features was 84, which took 2.66 seconds of CPU time to process the test set. When they reduced the number of features to 37, it only took 1.54 seconds to process the test set. While they only assessed network data (some classification errors were higher or lower depending on the attack type), they asserted that fusing from other heterogeneous sources such as a host log could be beneficial. Perhaps different feature selection techniques could have further reduced the number of features without significantly sacrificing classification accuracy.

Tsang et al. use a MOGFIDS (fuzzy-logic based) feature selection technique on KDD-Cup99 [38] and achieved the best overall feature selection results as compared to eleven other techniques in terms of classification accuracy. Chebrulu et al. [39] present an ensemble approach of feature selection and are able to achieve higher classification accuracies with the combination of feature selection techniques versus using each technique independently, and in their case they did improve overall Intrusion Detection accuracy for all attack categories and the normal class of the DARPA dataset by using an ensemble of Bayesian Networks and Classification and Regression Tress. Chen et al. [40] show that classification times can be sometimes be reduced in half when using SVM and C4.5 feature selection techniques on the KDD datasets, and they also evaluate Random Forest (RF) as a feature selection technique but they do not provide classification times for all the features of RF to compare its performance of classification times. Elngar et al. [41] use a Particle Swarm Optimization (PSO-Discretize-HNB) technique which uses feature selection to reduce the feature set size from 41 to 11 features, and with the smaller feature set Detection Accuracy improved from 97.7

Clearly feature selection can be beneficial with Intrusion Detection. However care must be taken in its application, as the nature of attack threats changes, so can the data. Correspondingly, the feature sets also change. Also similar to other domains, feature selection can be used to address Big Data challenges. Feature Selection can help reduce the dimensionality of the data being processed with Intrusion Detection, and it has the potential to mitigate Big Variety challenges simply through reduction of certain features. However, feature selection can also introduce additional Big Variety challenges when feature sets are highly unstable and a large Variety of different feature sets need to be utilized. Any time feature selection is used it can help address Big Volume challenges simply by collecting less data from the removal of certain features. Similarly, feature selection can be especially helpful in reducing Big Velocity challenges by increasing processing speeds. For example, Elngar et al. [41] found they could reduce processing times by a factor of ten simply by reducing 41 features to 11 features. Feature selection shows good promise for addressing Big Data challenges found within Intrusion Detection.

Using Hadoop to Ddress Big Data Challenges for Intrusion Detection

Traditional computing storage platforms like relational databases do not scale effectively against the onslaught of Big Data challenges posed by Intrusion Detection. Hadoop, an open-source distributed storage platform that can run on commodity hardware, has been utilized to better accommodate the Big Data storage requirements of massive Volume and fast Velocity along with potentially very diverse heterogeneous data structures. Collectively, Hadoop can refer to several technologies such as HDFS, Hive, MapReduce, Pig, etc. HDFS is the Hadoop Distributed File System, Hive is a data warehouse implementation for Hadoop, MapReduce is a programming model in Hadoop, and Pig is a querying language for Hadoop which has similarities to the SQL language for relational databases. Refer to [42] for further details on Hadoop.

Suthaharan [43] proposes the use of Big Data technologies like Hadoop, Hive, and the Cloud. He argues that before Big Data technologies should be employed to address Intrusion Detection, it should first be apparent that there are Big Data challenges present so as to not unnecessarily deploy Big Data technologies. Suthaharan argues that the current 3Vs of Volume, Variety, and Velocity cannot adequately provide for the early detection of Big Data, and so he proposes 3Cs of Cardinality, Continuity, and Complexity to more easily develop metrics with mathematical and statistical tools. A brief summary of the definitions for the proposed 3Cs follows: Cardinality - number of records at an instant Continuity - (1) continuous functions represent data; (2) continuous growth with respect to time Complexity - (1) data type variety is large; (2) high dimensionality; (3) high speed data processing

Suthaharan proposes a Big Data model to deal with Intrusion Detection as shown in Figure 2. The User Interaction and Learning System (UILS) performs the learning on the data, permits users to interact with the system, and can control the storage requirements. The Network Traffic Recording System (NTRS) simply captures the network traffic and either stores it locally in the Hadoop Distributed File System (HDFS) or the Cloud Computing Storage System (CCSS). If data is needed immediately it is stored locally in the HDFS, otherwise it can be stored in the CCSS and can be processed later. Also, for Machine Learning in Intrusion Detection and Big Data, Suthaharan recommends the following should receive more attention: multi-domain representation-learning, cross-domain representation-learning, and machine lifelong learning.

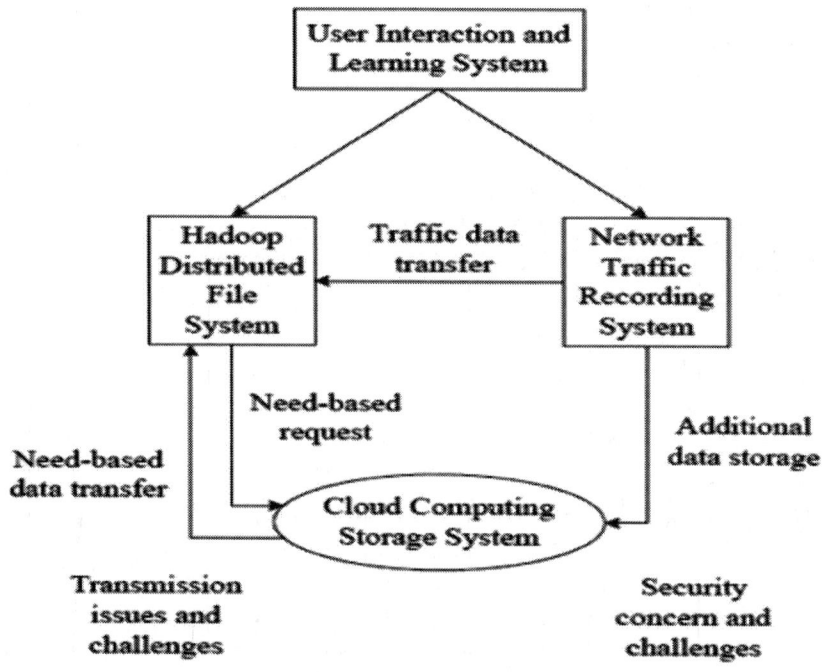

Figure 2: Suggested network topology for Big Data analytics [[43]].

Whitworth and Suthaharan [44] address the security challenges introduced with a model that can utilize the public Internet and the Cloud. Even though storage in the Cloud can incur a significant communication cost, higher latency, and additional security challenges, the authors contend that the Cloud can extend storage beyond a local network's capacity in an elastic and "cost effective and efficient manner" using Infrastructure as a Service (IaaS). Trust levels are proposed to assess varying levels of encryption requirements based on weighted values of cloud provider "risk level" and the sensitivity of the data. A Data Key Store (DKS) is also proposed to manage security and efficiently provide for data retrievability (ensuring the data is unchanged and available).

Jeong et al. [45] give an overview of issues encountered with Intrusion Detection and Big Data and how various Hadoop technologies can address these challenges, specifically focusing on anomaly-based (misuse) IDSs. They describe various techniques and issues found with Intrusion Detection, as well as what some of the main issues are in applying Hadoop technologies for Intrusion Detection. Their study provides a good introduction for readers not already familiar with Hadoop technologies and how they can be applied to Big Data challenges found with Intrusion Detection.

Lee and Lee [46] conducted an experiment with Hadoop technologies (e.g., HDFS, MapReduce, and Hive) to measure and analyze Internet traffic for a DDOS Detector. In their experiment they were able to achieve throughput speeds of up to 14 Gbps in some scenarios, and some of their slower results were close to 6 Gbps for some analysis types while using 30 or more nodes in a cluster. Several options were tested in the experiment, such as varying the number of cluster nodes (specifically, there were either 30 more powerful nodes or 300 less powerful nodes), and they also varied the file size of the playback file from 1 TB to 5 TB while performing 5 different types of analysis. Their study only considered previously recorded traffic data from files and not real-time traffic monitoring. However, they indicated that they plan to support real-time traffic monitoring with future work. Hadoop and its related technologies show good feasibility as an Intrusion Detection tool as they were able to achieve up to 14 Gbps for a DDOS detector, and this is only a preliminary experiment with future improvements planned.

Cheon and Choe [47] propose a distributed IDS architecture based on Snort and Hadoop technologies. They performed an experiment to see if additional Hadoop-based nodes for analysis could increase processing efficiency. Their methodology was to use replay files rather than real-time data, and then to evaluate the efficiency in terms of total processing time of the replay files while varying the number of Hadoop-based analysis nodes from zero to eight. A total of nine computers were used in the experiment with one acting as the "master node". They discovered that the performance efficiency increased (it took less time to process the dataset) as they increased the number of Hadoop-based nodes. However, processing efficiency actually decreased with only one Hadoop-based analysis node. When all eight nodes were used, they saw an increase of 424% in performance as compared to using the stand-alone machine without any distributed nodes of Hadoop analysis slaves. It would be interesting to repeat the experiment with significantly more Hadoop-based nodes in order to see how far this methodology can scale out and if a certain threshold would offer diminishing returns.

Veetil and Gao [48] conducted an experiment and created Hadoop clusters to implement the Naïve Bayes algorithm in a distributed fashion. With a 6 node "homogeneous" Hadoop-based cluster where the nodes had similar hardware, they were able perform classification 37% quicker than a stand-alone machine could. While the experiment was successful as a proof of concept to use a distributed Hadoop-based cluster to implement Naïve Bayes classifier, it could only classify an average of 434 packets per minute. Much more research and experimentation can be done to implement Hadoop technologies to improve Intrusion Detection efficiency and classification accuracy.

SECURITY DATA ACROSS HETEROGENEOUS SOURCES

The purpose of this section is to describe various techniques and architectures to accommodate diverse heterogeneous sources for Intrusion Detection. In order to not deviate from that focus since it is a central theme for this study, the Big Data implications of these systems will only be partially addressed within this section. Overall, Intrusion

Detection systems need to consider more diverse heterogeneous sources to provide better situational awareness within cyberspace. This can yield significant improvements to cybersecurity as Intrusion Detection is one of the core pillars of any cyber defense system. The first section gives a background on how data fusion can be used to improve situational awareness as has been done in other domains like Military applications. In the second section for illustrative purposes, a small sampling from academic studies of Intrusion Detection architectures with heterogeneous sources will be presented to give a brief overview and background of these systems. In the third section, several studies regarding SIEM systems will be presented, as well as some of the issues surrounding their deployment in the commercial sector. SIEM technology is not simply just another type of heterogeneous IDS architecture, but rather is a completely different architecture in its own right with an approach to heterogeneous data for Intrusion Detection which also provides for security analytics and forensic capabilities.

Enhancing Situational Awareness in Cyberspace with Data Fusion

In 2000, Bass [36] made a major contribution to Intrusion Detection research by suggesting data fusion as a technique to aggregate Intrusion Detection data from many different heterogeneous sources such as "numerous distributed packet sniffers, system log files, SNMP traps and queries, user profile databases, system messages, and operator commands". Essentially, data fusion is a technique to make overall sense of data from different sources which commonly have different data structures. Bass also elaborated extensively on using data fusion online (near real-time) in conjunction with data mining offline in order to process the enormous amount of cybersecurity data more effectively so that it could be useful for Intrusion Detection purposes. The purpose of the data mining is to discover previously undetected intrusions based on past data, and use these to build Intrusion Detection templates. This is not performed in real-time because the data mining operations cannot always be performed quickly enough to perform near real-time reactions for Intrusion Detection (which also suggests that Big Velocity was causing problems for real-time Intrusion Detection back in 2000). These Intrusion Detection templates are applied to the online (near real-time) data fusion operations in order to better assess possible threats.

Bass [36] described that he borrowed some concepts directly from military applications such as multisensor data fusion, where on the battlefield or in military theaters a widely diverse array of heterogeneous sources can be employed. He also described using a methodology discovered by the military's concept of Observe, Orient, Decide, and Act (OODA) to gain an overall higher cyberspace situational awareness by using data fusion for Intrusion Detection, and that data fusion can provide varying levels of inference from being merely aware of an intrusion attempt up to being able to analyze the threat and vulnerability. In "Multisensor data fusion for next generation distributed intrusion detection systems" [49], Bass elaborated further on his proposed model and provides further details on data fusion. Bass's approach of analyzing Intrusion Detection data across many different types of devices and systems concurrently is an excellent example of utilizing many diverse heterogeneous sources, helping researchers gain enhanced insight into cybersecurity (particularly in the context of Big Data challenges).

Similar to Bass's approach, Lan et al. [50] utilized data fusion across diverse heterogeneous sources with the explicit goal of improving Intrusion Detection through superior situational awareness. They warned that traditional deployments of security products such as Firewalls, IDSs, and security scanners rarely work together and only possess very minimal knowledge of the network assets they are protecting. In order to bolster cyber defense through a superior situational awareness, the authors proposed using a form of data fusion known as Dempster-Shafer (D-S) evidence theory in order to make good sense out of the heterogeneous sources. D-S evidence theory is a fairly common data fusion technique utilized by researchers using data fusion within the Intrusion Detection domain, which applies probabilistic techniques to the current observations of the system. Providing details for D-S evidence theory is beyond the scope of this study, so refer to [50] for more details.

A prevailing theme encountered by Lan et al. [50] was the Big Data challenges encountered when combining events from heterogeneous sources (e.g., IDS, firewall, host log files, netflow, etc.) to achieve better situational awareness. They discuss how Big Velocity problems can make it hard "to obtain the security state of the whole network precisely when facing too much warning information". Big Volume issues exist in collecting, fusing, and analyzing "a great deal of information". With diverse heterogeneous sources, the Big Variety challenges were very clear: "the complexities and diversities of security alert data on modern networks make such analysis extremely difficult". They conducted an experiment with the DARPA2000 data set, and by using data fusion were able to simplify alert messages from a total count of 64,481 to 6,164, which is an order of magnitude improvement. While this experiment shows data fusion can be an effective technique, further experiments using more modern and robust data sets would likely be of greater interest.

It is important to caution that just arbitrarily adding a multitude of sensors and fusing them all together does not necessarily improve accuracy. This is a phenomenon described by Mitchell [51] as "catastrophic fusion" where often the performance of an entire data fusion system is worse than that of the individual sensors. Careful design and consideration must be given to properly construct a data fusion system. Further background information regarding data fusion especially with regards to Intrusion Detection can be found in [52] by Hall.

This section described the importance of enhancing cyber defense through improving situational awareness. Just like data fusion is used in other domains for improving situational awareness, it can also be applied to Intrusion Detection. Research applying data fusion to Intrusion Detection shows good potential for improving the state of the art; however, researchers should carefully consider Big Data challenges that can exist within Intrusion Detection when applying data fusion.

A Sampling of Various Heterogeneous Intrusions Detection Architectures

The studies presented in this section give a brief conceptual overview of the various Intrusion Detection architectures found in academic studies when dealing with heterogeneous sources. Given that the previous section illustrated the importance of considering heterogeneous sources to improve cybersecurity, the purpose of this section is to explore the architectural issues of these systems identified by researchers. Following are five different examples of architectures proposed by researchers to accommodate diverse heterogeneous event sources.

In one study, Fessi et al. [53] consider Intrusion Detection across heterogeneous sources. A good illustration for this is given in Figure 3 where multiple distributed "Observers" harvest data from various heterogeneous sources (e.g., both network and various host-based monitoring) and a "Global Analyzer" makes the ultimate decision for whether security events originating from the "Observers" are security incidents. In making its final decision, this "Global Analyzer" will perform data fusion across the various "Analyzers" to gain a better situational awareness across the multiple analyzers especially in the case of distributed attacks. One of the interesting aspects of this model is that the "Analyzers" themselves can be heterogeneous, and different types of "Analyzers" such as misuse detection or anomaly detection could simultaneously be used for the same events from observers. So essentially, each observer could be associated to one or more "Analyzers" for the motivation of detecting different classes of attacks. This model could scale well in the face of Big Data challenges given some of its distributed characteristics, which enables "Observers" and "Analyzers" to be added for scalability. However if there is only one centralized "Global Analyzer", it could become a bottleneck in the face of very Big Data or it could also be problematic if it was successfully attacked or had reliability faults.

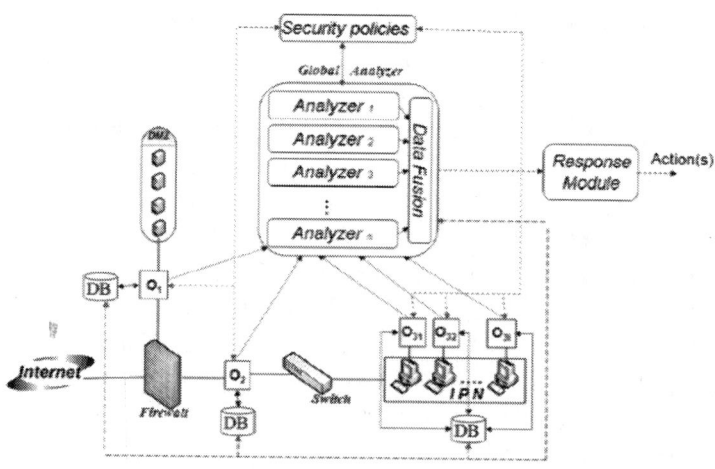

Figure 3: System Architecture [[53]].

To take more of a global view of Intrusion Detection, Ganame et al. [54] extend upon their earlier work with a centralized Security Operation Center (SOC) called a SOCBox, and develop an enhanced version called Distributed Security Operation Center (DSOC). Their architecture allows an organization to scale the system across the Internet to provide even better correlation across geographical boundaries and provide enhanced defense resiliency if one site comes under attack. Obviously, this architecture could even scale beyond multiple organizations.

One of the reasons Ganame et al. [54] extended the original centralized SOC architecture to DSOC was that the SOC architecture could be compromised by an attacker flooding the network on one site, and the Centralized SOC wouldn't be able to receive all the security events, allowing attackers to evade detection. They presented several examples of being able to compromise the SOC with "flood" attacks, demonstrating that the original SOC architecture was suspect to "flood" attacks and faced a Big Volume problem. The DSOC was able to overcome this problem by using a Local Analyzer (LA) at each site to assess intrusion detection by collecting, analyzing, and correlating alert security events locally. Each LA would then transmit a smaller and more intelligible payload of alerts to a Global Analyzer (GA) which would then perform further aggregation and analysis of alerts sent from all LAs in order to build better global awareness for Intrusion Detection (and the GA can be mirrored for redundancy and fault tolerance).

Ganame et al. [54] also describe the benefit of using diverse heterogeneous sources to correlate events across multiple sources in order to successfully detect attacks, and give an example where most homogeneous NIDS systems would be unable to detect certain multi-step attacks. Importantly, they were able to experimentally show that utilizing heterogeneous sources yielded superior Intrusion Detection capabilities over what most homogeneous approaches such as NIDS are capable of with more advanced attacks. The DSOC system utilizes diverse heterogeneous sources and accordingly monitors all network components such as "IDS, IPS, firewall, router, work-station, etc." to yield a more comprehensive situational awareness. Refer to Figure 4 for an illustration of examples that a Local Analyzer could use as diverse heterogeneous sources. The system also employs Protocol Agents and Application Agents to better facilitate harvesting the information from the source events in an understandable format as well as in a redundant fashion and with encrypted transmission. One other interesting aspect they discussed was the need for common message formats among different devices and protocols like the Intrusion Detection Message Exchange Format (IDMEF). However, they found that the XML bus used for IDMEF was "too heavy and resource consuming," especially for event correlation. The authors implemented a separate translation process to overcome this Big Velocity challenge.

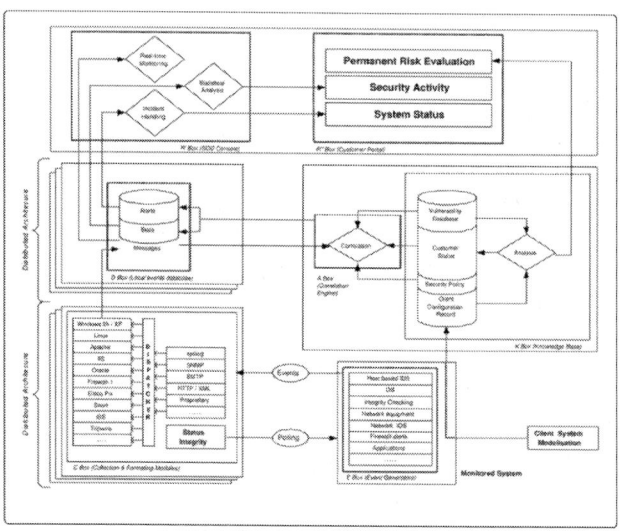

Figure 4: The internal architecture of a local analyzer [[54]].

This study demonstrates a couple of Big Volume challenges in that their original SOC architecture was prone to "flood" attacks, and that they could not directly use standard IDMEF formatting due to poor event correlation performance. Also, the use of heterogeneous data sources gave superior detection accuracy over homogeneous sources in some cases.

A Collaborative Intrusion Detection System (CIDS) is presented by Bye et al. in [55], where multiple "participants" (e.g., IDSs) form teams to work together to better assess Intrusion Detection collectively. As IDS technology has proliferated, the deployment of multiple IDSs within one environment has become more prevalent. A CIDS is a way for the multiple (and even different) IDSs to work together in teams. This allows a "Bigger Picture" to be realized through collaboration. The authors present a framework which can work across many different heterogeneous sources called Collaborative Intrusion Detection Framework (CIDF), and a set of mechanisms is used for a given detection or correlation algorithm to enable the Collaboration among IDSs. An "agent" is a participant in the CIDF which is also a member of a "detection group," possibly including other agents. These groups/subsets of agents have the same objective (such as anomaly detection), while another group/subset of agents may have a different objective (such as misuse detection). This study is relevant in that the authors are formally defining a framework for how CIDSs operate in general as well as how to cope with more complex issues such as security (and other issues) while collaborating. The authors also give examples of heterogeneous sources being used such as DSHIELD. Another interesting aspect is the overall heterogeneity of the framework, beyond just heterogeneous event sources. Agents within a group can themselves be comprised of heterogeneous "agents" (for example, by having different IDSs), and even the "detection groups" can be tasked with heterogeneous Intrusion Detection roles.

A Distributed Intrusion Detection System (DIDS) model is proposed by Bartos and Rehak [56] to overcome one major shortcoming of traditional IDSs: operating in isolation. Their main motivation is to increase overall accuracy and detect more threats. Importantly, their proposed DIDS can also accommodate heterogeneous sources, and their study gives examples of different event sources. The distributed IDS nodes are referred to as "sensors", and they have the capability to conduct data fusion to correlate different event types (i.e., if they are in different formats). Overall, every "sensor" IDS can communicate with every other sensor in the network with the motivation of redundancy as well as extra resiliency against attack. Each IDS "sensor" can tune itself to specialize its detection capabilities in order to improve accuracy for that specific attack class, and rely on other IDS "sensors" to evaluate other attack classes. Also, the IDS "sensors" can send requests for assistance when suspect behavior is encountered. Bartos and Rehak conducted an experiment for the proposed architecture and found that they could improve detection accuracy while keeping the false alarms constant. Their study is interesting in that data fusion across heterogeneous sources can help detection accuracy, but it is also interesting that it could not reduce false alarms especially considering that their architecture has a more global view. It would be interesting to evaluate the performance of this approach on a larger scale; however, it is a fairly novel exploration into utilizing distributed IDSs along with heterogeneous sources.

In evaluating DIDSs, Cai and Wu [57] discuss the software agent based approach for host-based systems where the agent monitors all relevant information of the host "including file system, logs and the kernel". While they also discuss the NIDS components, this is yet another example where more diverse heterogeneous sources are being monitored, enabling analysts "to get a broader view of what is occurring on their network as a whole". Cai and Wu also discuss the benefits of correlating IDS alerts across the Internet, similar to what Ganame et al. refer to in [54], and Bartos and Rehak also share this global view for Intrusion Detection in [56]. Other studies such as [58],[59] show alert correlation across geographical boundaries to be an important cyber defense strategy for the enterprise as a whole. In these studies, a prevailing theme is that more diverse heterogeneous sources will enhance Intrusion Detection capabilities through event correlation and a better comprehension of situational awareness of cyber threats.

Security Information and Event Management Systems

SIEM systems are architecturally different than typical IDS solutions, and are the result of computer security vendors in the commercial sector seeking to profit by solving problems that enterprises were experiencing. The SIEM term was first coined in 2005 by Gartner Analysts Mark Nicolett and Amrit Williams [60] to describe how the industry was converging Security Information Management (SIM) and Security Event Management (SEM) technologies. SEM primarily dealt with real-time analysis for the purposes of incident response, and SIM mostly dealt with the long-term storage for the purposes of historical and trend analysis as well as providing forensic capabilities. Anuar et al. [61] discuss additional background information on SIEM technology, specifically in terms of comparing SIEM products to more traditional IDS and IPS products. SIEM systems take a more comprehensive approach beyond traditional IDSs with the motivation of giving a better holistic view of an organization's IT security, and a good definition is given by Rouse [62] in that a SIEM gives the ability to see trends and patterns of security data from a single point of view even though the security data can originate from diverse heterogeneous sources such as the network, end-user devices, servers, firewalls, antivirus systems, and intrusion prevention systems.

According to Gartner [63] SIEM software sales was $976.4 million in 2012 with 27.5 percent growth, and for comparison the overall security software market grew from $17.7 billion in 2011 to $19.1 billion in 2012 as tracked by Gartner (with SIEM software comprising about 5% of the total market share in security software). Per Mosaic Security Research [64], there are currently 65 SIEM products as of this writing with 6 of them being classified as freeware. As SIEM technology is relatively new as compared to IDS technology, there are still many academic research opportunities especially considering the widespread commercial growth of SIEM technology. Following is a brief overview of SIEM technology. SIEM products can differ from each other in how they operate and in terms of features they provide, and one particular SIEM definition might not universally apply to all SIEM products. Aguirre and Alonso [65] generalize the major SIEM functionalities as the following: "aggregates data from many sources, continuously monitors incidents, correlates events, and issues alert notifications". In their study, they also contend it is important for organizations to aggregate SIEM information across their multiple domains and they propose a federation of SIEMs to accomplish that goal.

Similarly, after analyzing SIEM systems, Kotenko et al. [66] contend the four main SIEM components are the following: "event filtering, aggregation, abstraction, and correlation; reasoning and visualization; decision support reaction and counter measures; attack modeling and security evaluation".

Either definition is sufficient for SIEMs, especially since Kotenko et al. drew their definition from the most advanced SIEMs as defined by Gartner Analysts Nicolett and Kavanagh in [67], while Aguirre and Alonso apply their generalization to a broader range of SIEM products.

Kotenko et al. [66] also discuss the various standards used by SIEMs to represent security events and incidents in standardized formats: SCAP [68], Common Base Event (CBE) [69], and Common Information Model (CIM)[70]. One of the main design motivations of SIEM technology is that vendors will typically try to ensure all the security data sensor sources have as common a format as possible in order to minimize the amount of "data fusion" analysis that needs to be performed. They also discuss the inadequacy found in all the major and widespread SIEM systems in that they use a traditional relational database for the storage and querying of the security data, and that the relational database model that the SIEMs are using is often overloaded. To overcome this limitation, they recommend using a hybrid database approach for the SIEM repository where a traditional RDBMS is used in conjunction with both XML-based databases and a triplet store ("A triple store is a purpose-built database for the storage and retrieval of RDF metadata" [71] and the triple is based on the "subject-predicate-object" methodology). It would be interesting to determine whether Hadoop technologies could provide any storage repository benefits as that was not mentioned by Kotenko et al.

The fact that traditional RDBMSs are a performance bottleneck for SIEM systems demonstrates that they face major Big Data challenges. This makes sense as they are analyzing security data across a myriad of diverse heterogeneous sources, whereas this survey points out numerous times that Big Data challenges can be encountered at only single sources for Intrusion Detection. Kotenko and Chechulin [72] propose an interesting attack modeling framework for SIEM systems in [72] called Attack Modeling and Security Evaluation Component (AMSEC) to address both known and unknown (zero day) vulnerabilities.

Metzger et al. [59] conducted a study in Higher Education Institutes (HEIs) regarding Intrusion Detection and how it can be applied with SIEM technology in conjunction with formalized Incident Management techniques. The overall system can react to security events either automatically or manually through a Computer Security Incident Response Team (CSIRT). HEIs can be highly targeted among botnets, email spammers, and others for their high bandwidth capabilities among other reasons. The authors propose a framework where in addition to the traditional SIEM approach, a couple of other non-traditional sources for the SIEM system are considered: Manual Reporting and the "DFN-CERT service". Manual Reporting allows outside organizations or individuals, internal Administrators or Support staff, and Help Desk tickets to report security incidents and information directly to the system for automated processing. This extra source can benefit Intrusion Detection for the SIEM system with increased detection accuracy as it broadens the scope of events being monitored. The SIEM can also correlate its other events with this new source for increased benefit. The "DFN-CERT service" is a worldwide service to automatically report malicious behavior and metadata to the local SIEM system, with the similar benefit of enhanced detection and correlation capabilities for the SIEM. These two other methods are used in conjunction with the traditional SIEM monitoring, correlation, and analysis functionalities. Additionally, their model includes having the SIEM either take automatic responses to events or to notify appropriate Administrators for action based on configurable policies and/or threshold measurements of events. With their model, they were able to automatically react (at least partially) to more than 85% of all abuse cases in their HEI study. This is important in that it shows more heterogeneous sources can enhance detection, correlation, and reaction capabilities of SIEM systems, especially with regards to reporting more diverse security events and metadata to the local SIEM system. Benefit can also be gained with the local SIEM receiving cybersecurity intelligence from a worldwide network. Also, it is important to note that heterogeneous sources need not be limited to cyberspace as shown in this study, and that reports from humans can enhance the situational awareness as well. A major motivation for the system implementation was to automate everything as much as possible in a formalized manner.

In a study evaluating systems to monitor security for cloud computing, Diego et al. [73] conclude that no single solution can currently cover all existing security threats for Infrastructure as a Service (IaaS) platforms. To enhance the overall infrastructure security it is recommended to use more diverse and heterogeneous solutions. Diego et al. give an example that two different types of SIEMs could detect more threats than just one. In addition, this study proposes a Quality of Protection (QoP) in terms of both better fault tolerance and enhanced security for the security system itself by using systematic redundancy (i.e., if one part of the system fails or comes under attack then a redundant piece can still function). An experiment was carried out with the commercial ArcSight SIEM product to test the throughput when using redundant SIEMs in a Byzantine fault tolerant architecture. A total of 4 ArcSight SIEM "replicas" were used with one being allowed to be faulty, and over 250,000 events per second could be processed. They determined the system was bound by resource exhaustion, and additional resources could further increase event throughput. This study did not elaborate on the methodology of storage and querying of the events into the archival repository with regards to forensic purposes and how the archival repository would scale out as additional SIEMs were added to the system. This study is interesting from a Big Data standpoint in that it shows good experimental results in the scalability of SIEM technology, but there is no clear indication in whether SIEM technology would face a scalability threshold with relational databases still being the prevalent storage engine. However, recently some vendors such as Splunk [74] have adopted relational database technologies in order to better address Big Data challenges.

In an effort to make SIEM technology more effective in defending against rapidly evolving cyber threats, Li et al.[75] recommend an Enterprise Security Monitoring (ESM) solution. Large enterprises are facing increasingly challenging attacks such as Advanced Persistent Threats (APTs), and one major problem these large enterprises can have is their various security teams might be fragmented into different organizational silos which can cause difficulties in sharing security intelligence information across these boundaries to better correlate events, especially against more advanced attacks. Their proposal to advance cyber defense to face these challenges is to use SIEM technology as a core component, and use this in conjunction with Enterprise Security Intelligence (ESI) to enhance overall next generation cyber defense architectures. Essentially, ESI will extend the overall security intelligence of the SIEM capabilities similar to how Business Intelligence (BI) is traditionally applied, and would allow more advanced security intelligence analytics to be developed and utilized in order to adapt to more advanced threats. For example, to provide improved situational awareness with ESI, business context information specific to the organization could be combined with alerts generated from the SIEM as well as various intelligence sources (i.e., those reported by humans or systems).

As originally put forth by Gartner in [76], Li et al. explain the six design principals of the next generation ESM shown in Table 2 (Refer to Figure 5 for a visual illustration).

Table 2: Six design principals of the next generation ESM

1. "Comprehensive Enterprise Coverage"	The entire production IT stack (e.g., "networks, hosts, applications, databases, identities") for the enterprise must be monitored by the ESM regardless of environment (i.e., onsite or in the cloud).
2. "Information Interaction and Correlation"	All meaningful events, logs, and similar from input sources in #1 must be capable of being collected for correlation.
3. "Technology Interaction and Correlation"	The SIEM will serve as the foundation of the correlation engine, however it should also integrate with other important security technologies such as: Firewalls, IDSs/IPSs, DLPs, Vulnerability Management, and Anti-Malware.
4. "Business Interaction and Correlation"	The ESM must be aware and tuned to the specifics of the organization's business context to better assess an attacker's motivation and yield better correlation and intelligence.
5. "Cross-Boundary Intelligence for Better Decision Making"	The ESM solution must span organizational boundaries across the entire enterprise in a cohesive and collaborative manner, and not permit fragmentation with regards to its overall cyber defense.
6. "Visualized Output for Dynamic and Real-time Defense"	The output of the system must be easily visualized and understandable by end user analysts in an effective manner.

Zuech et al.

Zuech et al. Journal of Big Data 2015 2:3, doi:10.1186/s40537-015-0013-4

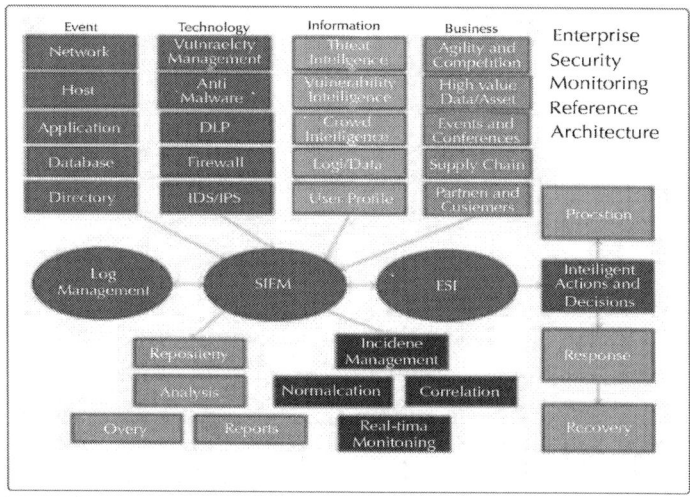

Figure 5: An enterprise security monitoring reference architecture [[75]].

The creation of a "fusion center" for the enterprise is recommended by the authors where the ESM is collaboratively utilized across organizational boundaries in a variety of ways especially regarding any aspects that could be fragmented (e.g., planning, risk assessment, data sources, intelligence analysis, etc.). They also contend that different organizations could even benefit by collaborating and sharing security information with each other. However they emphasize the great difficulty posed by this because of competitive, technical, legal, and possibly embarrassing reasons (i.e., disclosing certain breaches could harm their image). In order to better cope with the Big Data challenges of processing and storing "massive amounts of data", Li et al. suggest that technologies like Hadoop could be leveraged. They also recommend cloud-based Enterprise Security Monitoring vendors as a "natural solution" for Big Data and scalability issues of enterprises.

BIG HETEROGENEOUS DATA FOR INTRUSION DETECTION

The purpose of this section is to describe Big Heterogeneous Data in terms of different categories to illustrate the various underlying levels of heterogeneity for Big Data within Intrusion Detection. At a high level, Big Heterogeneous Data can be described in terms of being input or output data. Big Heterogeneous Input Data can be further categorized into traditional Big Cyberspace Data and Big Industrial Data (i.e., data from industrial processes in the real physical world). Big Heterogeneous Output Data will be presented in the categories of Big Archival Security Data (which considers the long term storage aspects) and Big Alert Data (which will present Big Data issues surrounding alert data).

Big Heterogeneous Input Data

Big Heterogeneous Input Data is essentially just the types of input data in the spirit of Big Heterogeneous Data from the previous section (security data across heterogeneous sources). It can be considered simply as just heterogeneous input Big Data. The following sections each discuss one particular type of heterogeneous input Big Data grouped by higher-level categories. It is important to consider that a great deal of heterogeneity among the sources can be present within these categories. First, the traditional cyberspace input Big Data is presented. Then, Big Heterogeneous Industrial Data beyond cyberspace is discussed, and this section gives examples of Big Data from the physical world outside of cyberspace (e.g., industrial process data) which can further improve situational awareness even in cyberspace.

Big Heterogeneous Cyberspace Data

Big Heterogeneous Cyberspace Data are the traditional input types of data which are commonly considered in Intrusion Detection literature, but here they are presented in the context of Big Data. Both network layer and host layer event sources are considered. The network layer coverage is essentially just the network traffic that traditional approaches like NIDSs (e.g., Snort) monitor with a focus on Big Data. The host layer coverage focuses on Big Data challenges with different host sources, and is equivalent to the traditional HIDS approaches where computer servers, workstations, devices, etc. are being monitored. Again, it is important to consider that a great deal of diverse heterogeneity can occur among event sources in this category.

Nassar et al. [1] contend that outsourcing flow-based network monitoring and Intrusion Detection to cloud providers can be cost effective if done so in a secure manner. They give an example of Big Data with a university network that produces an average load of 650 Mbps and peaks up to 1.0 Gbps, and assert that because of such Big Network Data that "many monitoring systems have already shifted from the deep packet inspection to the aggregated flow data level". In other words, because of such Big Velocity at the network level, more accurate techniques of intrusion detection such as deep packet inspection are being abandoned in favor of less computationally intensive techniques such as monitoring at the network flow data level. Nassar et al. discuss privacy and anonymization issues in being able to securely outsource network monitoring, and that a university network with an average load of 650 Mbps posing Big Velocity challenges is of particular interest.

In order to evaluate Big Network Data, Sitaram et al. [77] consider network-based IDS challenges faced by large cloud providers or those with fat network pipes such as OC-192 and OC-768 links. They consider such data as a "clear representation of big data streams in its most raw form (which is hundreds of thousands of TCP/IP packets per second)". Sitaram et al. envision building a NIDS capable of handling Big Data network streams such as these by utilizing Big Data tools such as Hadoop and a network monitoring tool called PacketPig [78]. According to the authors, PacketPig is capable of Deep Packet Inspection, deep network analysis, and even full packet capture when using it with Hadoop. In this study, they mainly consider the effectiveness of clustering algorithms for analyzing packet classification. Their experiment with the KDD data set found the K-means clustering algorithm to generally outperform the Expectation-Maximization and DBSCAN Clustering algorithms. However, their future work sounds especially interesting if it can successfully operate in terms of such Big Velocity.

Beyond the network, host-based event log data has traditionally been one of the main sources for Intrusion Detection monitoring. An organization can have a multitude of computing hosts both in quantity as well as diversity in terms of the different types of log files being generated. All of this log data can quickly add up to Big Host Event Log Data in that it can be very high in Volume, Velocity, and Variety.

The hosts that produce these logs can have Variety such as end-user computer workstations, computer infrastructure servers, devices, appliances, virtualization hosts, or even cloud-based hosts. The types of logs being generated are heterogeneous and can vary from Operating System events to a wide variety of application events such as antivirus software, firewall logs, honeypot activity, web server logs, ftp server logs, email server logs, domain controller logs, web proxy logs, VPN logs, DHCP server logs, etc. While this is not a comprehensive listing of the various types of logs one can encounter in the typical organization, it illustrates that the various types of logs can pose Big Variety challenges in having to correlate security events across a wide range of heterogeneous log types.

To better cope with Big Data challenges organizations can face with their log data, Yen et al. [79] developed a system called Beehive which performs "large-scale log analysis for detecting suspicious activity in enterprise networks". They report that organizations are facing Big Volume challenges in terms of the logs being "very large in volume", and implemented their system at a large enterprise, EMC, for two weeks. At EMC, they describe their major challenges as the "Big Data problem" where 1.4 billion log messages are generated on average per day (about 1 terabyte). This also suggests Big Velocity challenges in dealing with such a high data rate as well. They also discuss the problem of organizations implementing a variety of different security products which generate logs whose formats vary widely, and this suggests a Big Variety challenge. Also, they note that these logs from various security products may have problems such as incompleteness or even inconsistency, and so they describe logs with these challenges as being "dirty".

Beehive monitors the communication of dedicated hosts (e.g., workstations) with other host targets. This is accomplished by monitoring logs from a wide range of network devices such as web proxies, DHCP servers, VPN servers, windows domain controllers, and antivirus software. The logs are ultimately stored in a commercial SIEM system. They reported all the log information being stored in the SIEM as "a big data problem", and that "efficient data-reduction algorithms and techniques" are required to cope with such Big Data logging challenges. Another major challenge they encountered with the logs stored in the SIEM was that the information which was actually being stored proved difficult to correlate against other events because of the underlying quality of the data. As an example, logs could have the incorrect time stamp because of being in a different time zone. Another log format difficulty encountered was that typically only IP addresses were stored, and it was difficult to associate events to specific hosts given that IP addresses could change via DHCP. So, correlation against additional logs was necessary to properly identify hosts.

In terms of detecting security incidents, the Beehive project proved fairly successful versus the enterprise's state of the art security tools. Over a period of two weeks, 784 security incidents were discovered by Beehive whereas the enterprise's existing security tools only detected 8 of these incidents. So, a large number of security incidents were in fact unknown with the existing security tools. One source which was especially beneficial was the web proxy logs, where suspicious traffic activity or questionable destinations could be discovered. Yen et al.[79] were able to effectively reduce the amount of messages inspected by 74% simply by filtering out whitelisted target hosts. This is an interesting study in the effectiveness of the approach, especially considering the Big Data challenges encountered. It would be interesting to see if future similar work could perhaps extend beyond workstations to server log behavior.

According to Myers et al. [80], event correlation is frequently not performed with log analysis due to "difficulties and inadequacies with current technologies". One reason they indicate that organizations have difficulties analyzing security logs is because of "the sheer volume of data to collect, process and store". This suggests that log analysis with event correlation for Intrusion Detection is a Big Data challenge in the contexts of both volume and velocity. Myers et al. conducted an experiment on web server logs to evaluate the effectiveness of applying event correlation in a distributed fashion, and their results showed this technique could effectively detect many common web application attacks while maintaining a low false positive rate. Their distributed approach also showed a reduction in network traffic of syslog messages by 99.88%. This distributed approach illustrates good potential for addressing Big Velocity found in security network traffic by reducing that amount of traffic.

Big Heterogeneous Industrial Data

Cyber threats can damage and even destroy real-world physical targets beyond cyberspace. Industrial and Utility operations are especially prone to this exposure given their evolution of integrating and automating their physical operations with Information Technology from cyberspace. Even when these systems are "air gapped" and physically disconnected from the public Internet and other networks, these cyber threats can still be catastrophic in nature to real-world objects. An example of a successful attack occurred against Iran's nuclear program with the Stuxnet virus, and some of Iran's nuclear centrifuges were destroyed in the attack. Further details of this incident are given by Langner [81].

Therefore, it can also be important to include heterogeneous sources from the physical world to better improve overall situational awareness for security. A good conceptual illustration for how to extend monitoring beyond cyberspace is given in Figure 6, and this shows different Host, Network, and Device IDSs harvesting information into a centralized SIEM system with the goal of improving Intrusion Detection by also analyzing data from Process Control System sensors. This illustration is from a study performed by Valdes and Cheung [82] with the explicit goal of gaining better situational awareness in process control systems. The motivation was to extend the existing functionality of a SIEM product (ArcSight) to correlate control/process system alarms with IDS events from the SIEM, and thus extend situational awareness beyond cyberspace to also include industrial physical process control systems by correlating IDS data with measurements from underlying physical process data such as electrical current, pressure, flow rate, and similar industrial measurements. This is an interesting concept in that cyberspace situational awareness can be improved by correlating data from heterogeneous sources in the physical world beyond cyberspace, and that Intrusion Detection need not be merely limited to cyberspace sources. The authors indicate that important industries such as refining, pipelines, and electric power can benefit from this approach of utilizing more diverse heterogeneous sources, while cautioning that the stakes are especially high for detecting cyber-attacks against those platforms, as damage can also be physically harmful or even deadly, such as releasing hazardous materials into the environment. Refer to [83] for further background information and also the final report [84].

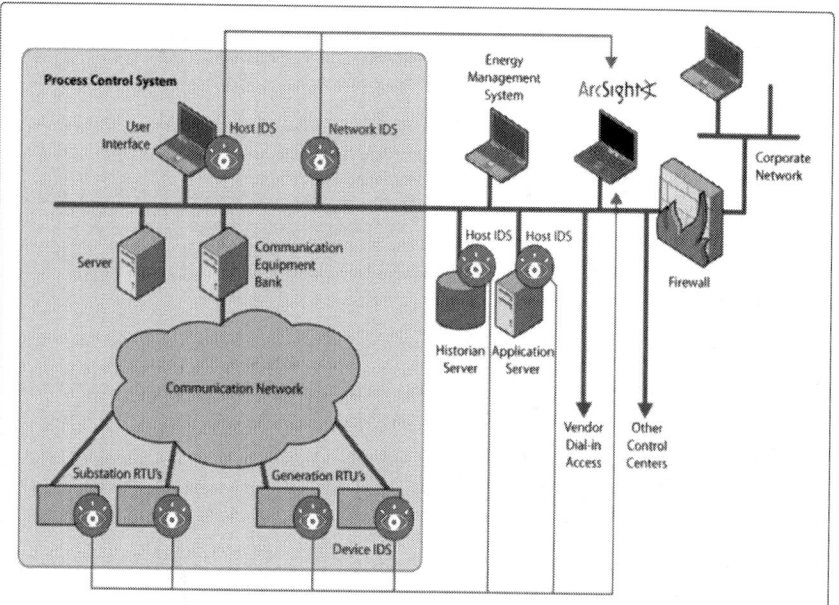

Figure 6: Control Center Level View of the Detection and Event Correlation Frame-work [[82]].

According to Xiao-bo et al. in [85], Data from Industrial and Utility operations certainly can have Big Data. They design "a big data acquisition engine based on rule engine" to better handle Big Data acquisition flow problems when industrial processes face such challenges. Their study describes how supervisory control and data acquisition (SCADA) systems are evolving and increasingly using Ethernet-based networks rather than traditional serial port connections. SCADA systems are essentially Industrial Control Systems which are based on computers, and they control and monitor industrial processes in the physical world. Xiao-bo et al. propose a rule engine implemented in Java Expert System Shell (JESS), and seek to improve performance and quality from industrial Big Data acquisition flows. Their design accommodates the real-time demands of SCADA systems, and can be used for data analysis, alarming, and forecasting. It is interesting that they chose to address Big Data challenges as a cornerstone of their design in dealing with data flows from SCADA systems.

A study conducted by Datta Ray et al. [86] evaluates Security and Big Data challenges that the electric power grid is facing. They assert that a paradigm shift is required to properly address the cybersecurity demands of the smart grid (i.e., electric power grid). Even more diverse heterogeneous event sources will need to be monitored, and this will further exacerbate the "Big Data onslaught" the utilities are currently facing. Additionally, a Formal Risk Management system should make "the analysis of such Big Data manageable, scalable, and effective." A more formal Return On Investment (ROI) analysis should rigorously address an enterprise's multitude of security and risk contexts.

When it comes to cybersecurity, Datta Ray et al. contend that "the crux of the problem is that organizations have taken a piecemeal approach to security". Various security products such as firewalls and antivirus software do not communicate systematically with each other to yield "holistic intelligence", and typically by the time meaningful patterns are found it is too late and the damage has already occurred. They further assert "these point or perimeter solutions applied to host computers, networks, or applications often work with little knowledge of each other's functions and capabilities". In order to be successful in achieving a holistic risk management system, a major design consideration is interoperability between a diverse myriad of devices such as "meters, synchrophasors, IEDs, firewalls, field devices, etc." This interoperability is important, and should even consider both structured and unstructured data from the sources. By interoperating among "existing point, perimeter, and defense-in-depth security solutions with actionable insights", a more systematic and superior cyber defense can be realized.

In this spirit of considering more diverse heterogeneous security event sources, Datta Ray et al. provide great detail on enhancing overall security by integrating security event sources beyond cyberspace and the Big Data challenges of doing so. The model that they use to illustrate this is by categorizing traditional cyberspace event sources as Utility Business Information Technology (IT), and by categorizing event sources beyond cyberspace as Power System Operation Technologies (OT). For example, IT event sources could be typical cyberspace components such as firewalls, and OT event sources could be a meter that measures electrical quantities such as power, voltage, and current.

By combining and correlating security events across both IT and OT sources, an improved situational awareness can be realized. However, both the IT or OT sources can face Big Data challenges, and Datta Ray et al. propose a model shown in Figure 7 on how to cope with the Big Data onslaught in a systemic manner to improve risk management. In addition to diverse heterogeneous sources, the model also considers "exogenous and endogenous sources of intelligence and asynchronous and real time interactions among its various components". The model indicates how the system can provide feedback in near real time to react to specific events, and that the intelligence of the system is based on an aggregate of Big Data IT and OT inputs.

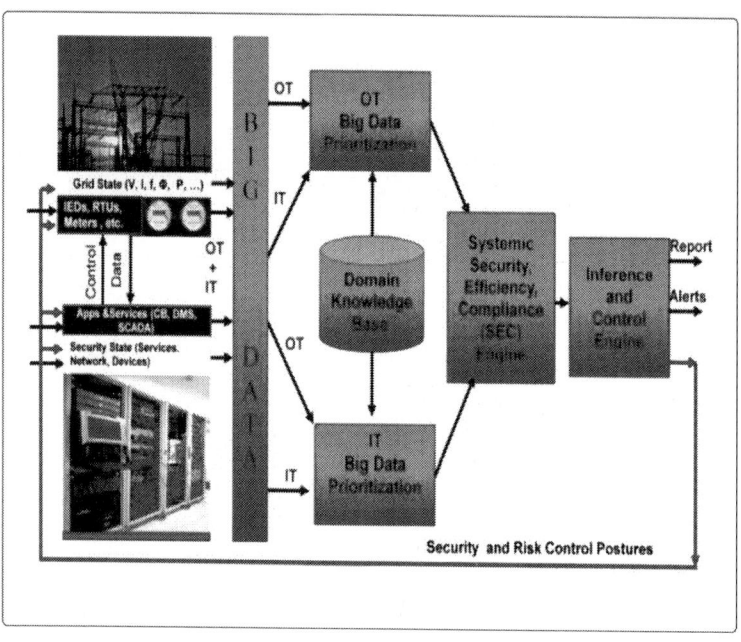

Figure 7: Big Data and OT, IT Systemic Risk Management [[86]].

By unifying the IT and OT domains and correlating events across the entire spectrum, the overall quality of the entire smart grid can be enhanced even while processing a Big Velocity of information from a Big Variety of diverse sources which can lead to valuable insights from previously unknown correlations and hidden patterns. The smart grid faces all of the "3Vs" of Big Data criteria: "huge volume, diverse sources and types and the varied speed of the incoming data". The amount of data the smart grid faces is so large that it has "surpassed the ability of traditional relational and scan/sort systems to process the data".

Formal Risk Management and ROI analyses are very important considerations in how utility companies process Big Data, as the risks are high. While financial risks such as electrical theft or outages are obvious, the loss of human life is possible in extreme cases like a gas explosion from a leak, or if a hospital (or home) that uses life-support systems were to be maliciously denied electricity. Instead of merely "drowning in data", Big Data can be used as an advantage to improve risk analysis and combined with ROI analysis can more effectively prioritize information assets in terms of effectiveness, scalability, and protection.

Big Heterogeneous Output Data

Big Heterogeneous Data can be output data as well, and this is classified as Big Heterogeneous Output Data. This section addresses the heterogeneity of output Big Data for Intrusion Detection in two main categories: Big Archival Data and Big Alert Data. Big Archival Security Data is output data which is being archived either for the purpose of forensics or Security Analytics, while Big Alert Data is output data either for further alerting analysis or for notifying an administrator or system component to take action. Both of these Big Heterogeneous Outputs can have very pronounced Big Data attributes in terms of Volume, Velocity, and Variety.

Big Archival Security Data

A very important aspect for Intrusion Detection is long-term storage of certain security data. Essentially, there are two main goals for the archival of security data. The first goal is to improve Intrusion Detection capabilities even in real-time with offline data mining operations and Security Analytics. This offline data mining operation on security data can further try to identify previously unknown cyber threats, and then update the real-time detection capabilities with additional new signatures or behavior traits. The second goal is to provide forensic capabilities with this data so that in the event of a security breach, forensic evidence is available to assist the investigation. This data can also be used as evidence in legal proceedings if properly maintained. Typically not every single piece of computing data will be kept in the offline repository store, and care must be taken to properly filter out what is not necessary.

In an experiment using log data generated from anti-malware software, Hoppe et al. [87] use data mining to search for patterns among malware infections in the archived storage repository of a SIEM. They described difficulties in performing the data mining in dealing with such a large amount of data, and that a critical success factor was utilizing the formal CRISP-DM data mining process. In addition, they also described that in IS infrastructures: "the amount of data is enormous". Big Data challenges were a factor in their study especially in the context of Big Volume, given that their data mining operations were performed offline and not for real-time Intrusion Detection. In their study they found that the age and gender of workstation users could infer high or low risk of malware infection. However it was also found that users with or without administrator rights on their workstations did not influence malware infection. Hoppe et al. contend that in specific scenarios, performing data mining on data collected by SIEM systems can enhance the quality of Information Security Infrastructure for companies. This is an illustration that feedback from the offline archive store of a SIEM can be useful for better real-time inference of events which might even possibly yield better situational awareness.

A model proposed by Hunt et al. [88] seeks to both enhance real-time adaptive security while improving long-term forensic capabilities. They point out that security data "arriving too fast to store" and process can now be better addressed with new terabyte storage devices, parallel processing, clustered computers, and even super computers. To give a point of reference, a network with a 10Gbps flow over one hour of incoming traffic requires 5 terabytes of storage. They assert that such infrastructure is currently out of reach for medium sized organizations. Given the processing and storage problems and the application of super computers, there are clear Big Data challenges, especially in the context of Volume and Velocity.

According to the Hunt et al., most IDS/IPS and firewall systems even when reporting information to SIEM systems frequently do not capture sufficient information for robust forensic capabilities as they do not create "digital evidence bags". These systems usually do not sufficiently automatically react in real-time or provide sufficient "traceback" functionality. "Traceback" functionality is the ability to correlate already identified malicious sources (e.g., source IP, port, ISP, etc.) with other real-time components of the network as well as for future forensic purposes. In their model, they also suggest that honeypots, honeynets, and sinkholes are important components for the overall system. Sinkholes can be summarized as a way to draw these packets into a "sinkhole" and allow the malicious packets in so that they can be recorded for forensics as well as later correlation of suspicious behavior, versus having a firewall merely dropping packets with malicious characteristics. These malicious packets are ultimately drowned and not permitted to propagate past the sinkhole, as another firewall explicitly blocks these packets.

Hunt et al. also assert that Data Loss Prevention (DLP) systems are mature in some regards. However their capability to link back to the original events pertaining to breaches with forensics is "largely lacking". They emphasize that DLPs should integrate better with SIEM systems to improve the "very serious" situation of needing better forensic capabilities. Importantly they realize it is "inevitable" that for both real-time security and forensics the focus needs to shift from network protection to data centric protection (i.e., data leakage from the database). Intuitively, this is a compelling argument given that gaining access to such data is a major motivation for attackers.

While their model gives many interesting details, a few examples will be briefly mentioned. One of the recommendations is to use encryption for the log devices from all sources, and to use a "digital evidence bag" for the purposes of forensically sound data (including the possibility of using a kernel security module to mitigate interception attacks). Also emphasized is that a proper chain of custody must be employed in order to be able to properly prosecute perpetrators, and a couple of helpful items for this would be cryptographic hashes and key management for all evidence.

Their proposed model seeks to improve upon existing forensic capabilities while also enhancing real-time Intrusion Detection with additional correlation sources. The authors emphasize that not all systems have these weaknesses, although many do. They also indicate that the extent to which correlated data can adapt firewall rules in the real-time is an open research question.

Big Alert Data

Intrusion Detection Systems and other security systems produce alerts to notify administrators of suspicious activity. Even an individual IDS can trigger many alerts, and the problem becomes even more prominent when dealing with heterogeneous sources such as a wide array of sensors or multiple IDSs. The basic problem is that a single security inspection event can trigger many alarms even if it is a single incident, or many false alarms can even be raised with normal traffic.

A common technique which is used to stop a flood of alerts is called alert correlation. The basic concept of alert correlation is that when the same characteristic is causing the same alarm, the system should filter and aggregate multiple alarms into one alarm so that a flood of alarms of the same type does not occur (instead just a count of those same alarm types could be reported). An illustrative example of alert correlation is given in Figure 8 where alerts are initially correlated locally in a hierarchical fashion. They are subsequently correlated again at a more global level.

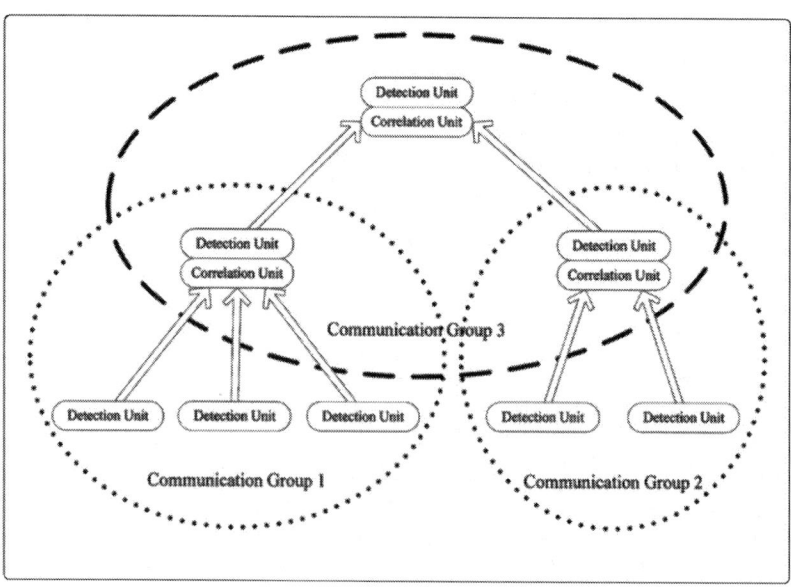

Figure 8: Hierarchical CIDS architecture [[9]].

The process of generating alerts certainly can involve Big Data challenges in terms of Volume, Velocity, and Variety. Big Volume and Big Velocity challenges for alert generation can involve correlation with other alerts, events, rules, or knowledge bases. These correlation activities can involve massive processing power, storage requirements, and network traffic. Big Variety challenges for alert generation can involve correlation among alert generators such as IDSs that can have many different formats for their alert messages or event data. It is common for organizations to have security products with many different proprietary alert formats, even though efforts are still being made to standardize. Semantically, alerts can either be considered inputs or outputs as they can also serve as inputs for alert correlation purposes. Alerts always operate at least once in an output capacity, but alerts do not always operate in an input capacity. Since alerts are typically considered outputs conceptually for notification purposes as well as for archiving and forensic purposes, they will be categorized as outputs for this study's organizational purposes.

The actual alerts themselves can indeed pose Big Data challenges, and a study by Sundaramurthy et al. [89] gives an example where they analyzed a data set with over 35 billion alerts from the HP TippingPoint IPS product. This data was collected over a 5 year period in over 1,000 customer networks worldwide. The data mining analytics posed the most significant Big Data challenges given the sheer volume and processing requirements which included correlating the alert data with the filter metadata (which contains further details about the actual alert). One interesting insight learned from the data was that with the Denial of Service (DOS) or Distributed Denial of Service (DDOS) attacks, the majority of these attacks were actually attacking the application layer instead of the network layer. Traditionally, DOS attacks are at the network layer where a bunch of packets "chokes" a device. However, newer attacks can more effectively attack the application layer by sending a specially crafted malicious packet which will cause the application to consume significant resources. They found 78.65% of the DOS alerts were from application-layer attacks, and only 21.35% of the DOS alerts were from network-layer attacks. DOS attacks can present Big Data challenges in themselves. If alerts from these DOS attacks are not properly handled via alert correlation or similar techniques, then the alerts themselves can actually compound the Big Data problems from the DOS attacks with alert floods. These DOS attacks pose a Big

Velocity problem, and their corresponding alerts could as well if not properly handled.

For Collaborative Intrusion Detection Systems (CIDSs), Zhou et al. [58] found that a decentralized approach for alert correlation versus a centralized approach could significantly reduce the processing time and number of alarms while not significantly sacrificing detection accuracy. This is especially important given that one of their major challenges was a Big Data challenge with the alert messages themselves being bottlenecked at a centralized alert correlation server: "the scalability problem addresses the challenge of how to cope with the huge volume of raw alerts that can be generated by each participating IDS in the system". They were able to solve this bottleneck problem by utilizing a hierarchical alert correlation architecture as shown in Figure 8. For a given stealthy scan scenario, they were able to significantly increase detection accuracy with their probabilistic threshold approach versus a naïve scheme that used the same threshold. This architecture was able to accommodate multi-dimensional data sources by restricting their analysis to only four specific features (Source IP Address, Source Port, Destination Port, and Protocol), and Zhou et al. applied these four features to evaluate eight different combinations containing them. Admittedly their approach had drawbacks such as not being able to always properly detect if the attacker spread their attack over a longer period of time than the threshold could handle. Future work for their architecture should also address security countermeasures if distributed node(s) are compromised, which is quite possible considering that this architecture is a distributed system that could span across the public Internet. Their study is interesting from a Big Data perspective in that they could significantly improve overall detection time by actually benefiting from the increased message routing and processing time in a distributed architecture. However, their study obviously only considered network traffic features and was not a comprehensive evaluation across heterogeneous sources beyond network traffic features. More work in this area should evaluate just that.

The processing for alert correlation is clearly a Big Data challenge. Roschke et al. conducted an experiment [90] to evaluate the efficiency of alert correlation with three different database architectures: traditional row-based DBMS (e.g., MySQL), memory-based column database (e.g., MonetDB), and a custom built in-memory database. In all cases, the in-memory database yielded far superior processing for alert correlation with the authors concluding that the row-based database showed "poor performance" for clustering and correlation. While the column database performed better than the row-based database in terms of alert clustering and correlation analysis, the column database performed extremely poor in terms of record insertion, being able to only insert 63 records per second, and so the authors concluded that particular database was "unfeasible" for general IDS utility and only useful for analytic purposes. The authors do point out that the utility of the in-memory database is obviously limited by the amount of memory available which will give an upper limit to the amount of alerts that can be correlated (or clustered), whereas traditional databases do not have this limitation. They propose limiting the time window in which alerts can be correlated or clustered to avoid this problem. This experiment shows that alert correlation is a Big Velocity challenge, although this experiment was limited to only evaluating network-based alerts via Snort. Big Data challenges for alert correlation would be further exacerbated if additional diverse heterogeneous sources were utilized as the number of alerts could increase, in addition to the broader need to perform data fusion across the different sources.

DISCUSSION

When it comes to Intrusion Detection for Big Heterogeneous Data, a great deal of future work can be done with research. If it can be presumed that the annual cost in the USA for cybercrime is $100 Billion annually [17] and that organizations spend about 26% on Detection [19], obviously it cannot be inferred that Intrusion Detection is costing $26 Billion annually (as the total national cost for Intrusion Detection cannot be directly extrapolated from amounts spent by organizations for Intrusion Detection and national Cybercrime costs). However, even if these numbers are only estimates it can be gauged in terms of the order of magnitude that the national costs for Intrusion Detection are very large. Analyzing much more heterogeneous security data can yield significant improvements to the Intrusion Detection domain, and employing Big Data technologies and techniques will allow more of this Big Heterogeneous Data to be utilized. Promising future research in this spirit will be presented in the following main areas: Data Sets, Feature Selection, Data Fusion, SIEMs, Database Issues, and Other Architectural Considerations.

Data Sets

Clearly, the relative lack of high-quality Intrusion Detection data sets is a problem for the academic research community. Some researchers have been building their own data sets due to the lack of existing relevant ones. However one difficulty with this approach is how to accurately label Intrusion Detection data as either attack or normal. Honeypots, honeynets, and sinkholes as well as intentionally cyber-attacking for the purposes of generating data sets might be able to help with labeling data as an attack, but there are still challenges with these approaches. Also, other traffic cannot always be assumed to be normal as it could also be contaminated with attack data. Synthesizing both attack and normal data might not build robust enough datasets in that they are not similar enough to the real world. Another issue is that data sets should constantly be updated as time progresses with new instances containing new normal traffic (i.e., new technologies, applications, and users) and attacks (i.e., new techniques or exploits) to keep research relevant as well as to better train learning systems for Intrusion Detection in an iterative fashion as technology and cyber-attacks evolve.

High quality, robust, and heterogeneously diverse public data sets are so fundamental to Intrusion Detection experimentation and research that the availability of such data sets is fundamentally a significant research problem in its own right. Actual research into the production of these data sets could allow researchers to make better overall progress in the Intrusion Detection arms race. Various techniques and issues such as those mentioned in the preceding paragraph and in the Intrusion Detection Data Set Challenges section should be considered when developing such data sets. Another important consideration is that Intrusion Detection data sets should also accommodate many diverse heterogeneous security sources to adequately address the issue of improving situational awareness. In addition, perhaps Intrusion Detection may need to use more than binary classification in some scenarios and use multiple classifications (i.e., "attack", "normal", "suspicious", "unknown", etc.).

Feature Selection

The application of feature selection can be helpful for real-time intrusion detection in terms of reducing detection classification times and even sometimes improving classification accuracy, but care must be taken in the application of feature selection. It is important to understand where and how feature selection can be applied effectively, especially with regards to unstable Intrusion Detection data sets. More specifically, how can feature selection play a role in security event data that is constantly evolving with new characteristics due to issues such as new technologies as well as new cyber-attacks? Thus, the stability of feature sets is very significant in the cybersecurity domain where the landscape is extremely dynamic and not as static as some other domains (e.g., bioinformatics) where feature selection is applied. Feature selection could play a very prominent role where many diverse heterogeneous sources are being analyzed and correlated as feature reduction could significantly reduce storage Volume, processing Velocity, and complex Variety.

Data Fusion

Data Fusion has not been widely adopted within the Intrusion Detection domain as compared to other domains like military applications that have a multitude of diverse heterogeneous sensors. A considerable amount of research has been conducted for alert correlation with Intrusion Detection, but very few works consider event fusion or other types of data fusion. Many more experiments with different data fusion techniques should be performed, especially in the context of many more diverse heterogeneous sources which contain Big Data. The question is therefore raised: can significant improvements be realized for Intrusion Detection by doing this, or would the costs be too high?

SIEMs

SIEM technology is proprietary and it is difficult to speculate on some of the internals of different product offerings, but it is apparent that SIEMs do attempt to normalize the data sources into as common formats as possible (e.g., SCAP, CBE, and CEE). While standardizing the formats of security events is good in the sense of reducing some of the more challenging data fusion aspects and to increase performance, in some cases it could possibly be detrimental if valuable information for a particular source is being lost or minimized. However, standardized message formats are not a reality in the real-world in many situations due to a wide variety of security products simply still using different formats. More experimentation into how SIEMs compare to traditional data fusion approaches might be beneficial. Obviously there are benefits to formatting all security event sources into as common formats as possible. However what are the drawbacks with the common format approach from actual experiments?

NIDSs and HIDSs still play a prominent role in the commercial sector. However SIEMs seem to have recently taken the security industry by storm for larger organizations with their ability to both detect cyber-attacks in the real-time, and provide offline security analytics and forensics. But the research community still seems too narrowly focused on the Intrusion Detection problem by conducting much more research on traditional IDSs than a more heterogeneous approach such as SIEMs. SIEMs should be evaluated much more by the research community, especially with regards to experimentation. Furthermore, the research community could further evaluate how to extend SIEM functionality in novel ways as SIEMs have not been Intrusion Detection's "silver bullet".

Database Issues

When Big Data poses challenges for Intrusion Detection, whether it is in standalone subsystem event sources or major architectural components for a heterogeneous system, how can technologies like Hadoop and similar technologies improve upon the state of the art and solve pressing problems? More experimentation is needed in this regard. Specifically for SIEMs (or their research-based data fusion equivalents), how can we improve the efficiency of the offline repository stores (which are used for both forensics as well as offline security analytics which provide feedback to the online system?) Many SIEMs use traditional RDBMS technology for this purpose, and more experiments could be conducted to see how other approaches such as columnar databases, xml databases, Hadoop technologies, or hybrids thereof could make these repository stores more efficient and effective.

Research into using better storage platforms effectively is needed for the enormous Volume, fast Velocity, and complex Variety processing requirements for Intrusion Detection.

Other Architectural Considerations

From a general conceptual framework, more experimentation is needed into what Bass [36] proposed, which explored how data mining from offline repositories can give useful feedback to effectively and efficiently benefit real-time Intrusion Detection. Bass's concept was fairly sophisticated and very highly cited by the research community. However there has not been significant experimental research with this model. A good deal of experimentation is still needed to explore the effectiveness of Bass's model with concepts such as real-time feedback and utilizing different Intrusion Detection "feature templates" based on the current situation. Similarly, SIEMs do afford archive repository storage for security analytics and forensics, and researchers could evaluate how to further improve the real-time Intrusion Detection component with this repository store.

In general, more experimentation is needed to illustrate the effectiveness of the cloud in solving Big Heterogeneous Data challenges for Intrusion Detection. More studies and experiments to illustrate the costs involved with utilizing the cloud platform would be beneficial in terms of network bandwidth, processing, and storage requirements. An important consideration is to determine what data gets filtered out, which is an important consequence in terms of establishing good ground truth for Intrusion Detection as well as regarding forensic capabilities.

Some good preliminary research has been conducted regarding architectural topology in terms of Centralized, Distributed, or Hybrid approaches for Sensors, IDSs, and Analyzers. Typically SIEMs will use a Centralized approach for the decision unit which possesses advantages and disadvantages. More experiments could be conducted to further refine these. Also, more research into how SIEM systems can scale out with multiple SIEMs would be beneficial as well. Some benefits have been realized with distributed or hierarchical architectures for IDSs and analyzers, but much more potential experimentation is yet to be realized especially in terms of utilizing more data fusion.

Heterogeneity among the actual Sensors, IDSs, Analyzers, or even SIEMs can be beneficial for Intrusion Detection where detection accuracy can be improved. For example, different IDSs working in teams to evaluate the same security events can improve detection accuracy, and the same can be accomplished with using more than one type of SIEM to analyze the same security events. So far, this type of Detector heterogeneity has showed good benefits. Even more research should be conducted into this area especially in terms of being sensitive to the additional costs versus the benefits.

Greater geographical and organizational heterogeneity should be employed for Intrusion Detection. Alert and event correlation beyond geographical and organizational boundaries could further improve situational awareness. Projects such as the Internet Storm Center [34] for honeypots and collecting malicious Intrusion Detection metadata. However organizations could participate much more actively. Will a natural evolution be to outsource the Intrusion Detection function to Managed Service Providers (MSPs)? Will these MSPs be able to more cost effectively manage Intrusion Detection through scaling out with a more comprehensive Intrusion Detection knowledge base and better core competency? More research should evaluate how significantly sharing of Intrusion Detection events, alerts, analysis, and knowledge across many organizations could enhance the state of the art. Are there ways for even competing organizations to share cyber-threat security event data for their collective good while still protecting their competitive interests? A prevalent consensus among the research community is that greater sharing for Intrusion Detection is necessary, and more research should be conducted to evaluate this on a significantly larger scale of sharing.

CONCLUSIONS

Historically most of the academic research for Intrusion Detection has focused too narrowly on the network layer with NIDs and to some extent at the host level with HIDSs. The academic research community should actively embrace a more diverse heterogeneous-based event source approach and follow the lead of the commercial sector where the rapid proliferation of SIEM technology has blossomed into a billion dollar industry (despite the term "SIEMS" having only been coined in 2005). This proliferation of SIEM technology throughout industry is an important consideration, given that one of the inherent features in this technology is to correlate security events from a wide array of diverse heterogeneous sources, and its successes in the commercial sector should give credence to this approach.

Both cybersecurity and physical security for organizations such as those in the utility and the industrial sector can even be enhanced by correlating traditional IT security events with those beyond cyberspace such as sensor devices measuring anomalous real-world quantities like gas leaks, electrical power/voltage/current, temperature, fire alarms, or many other sensors. Correlating security events from physical world sensors with cyberspace is becoming significantly more important as the utility and industrial sectors are becoming increasingly computerized for automation, and thus exposing their physical infrastructures to new cyber threats such as malicious attackers or "cyber accidents".

More diverse heterogeneous sources can provide for improved situational awareness within the Intrusion Detection domain similar to the military's use of diverse heterogeneous sources in its doctrines, strategies, tactics, and engagements. The onslaught of all this Big Input Data drives the engine for an onslaught of Big Output Data. For Intrusion Detection, great heterogeneous diversity in both its input and output data poses significant Big Heterogeneous Data challenges.

While Intrusion Detection does not always face Big Data challenges, it does face Big Data challenges more often as time progresses and especially more so for larger private and government organizations. This trend of Big Data challenges will continue as a multitude of more heterogeneous sources are analyzed. Even medium and smaller organizations will need to assess whether their Intrusion Detection architecture or Security Analytics merit the deployment costs of Big Data technologies. Big Data challenges for Intrusion Detection already exist for the nation's electric grid given its tight integration with computers, and will become even more pronounced as many more diverse cyber and non-cyber heterogeneous sources are brought online to enhance overall cyber defense and improve situational awareness for critical infrastructure.

AUTHORS' CONTRIBUTIONS

RZ performed the primary literature review and analysis for this work, and also drafted the manuscript. RW worked with RZ to develop the article's framework and focus. TMK introduced this topic to RZ and RW, and coordinated the other authors to complete and finalize this work. All authors read and approved the final manuscript.

ACKNOWLEDGEMENTS

We thank the reviewers for their constructive comments.

REFERENCES

1.   Nassar M, al Bouna B, Malluhi Q: Secure outsourcing of network flow data analysis. In *Big Data (BigData Congress), 2013 IEEE International Congress On*. IEEE, Santa Clara, CA, USA; 2013:431-432.

2.   Group BDW (2013) Big Data Analytics for Security Intelligence. https://downloads.cloudsecurityalliance.org/initiatives/bdwg/Big_Data_Analytics_for_ Security_Intelligence.pdf. Accessed 2015-1-10.

3.   Chickowski E (2012) A Case Study In Security Big Data Analysis. http://www.darkreading.com/analytics/security-monitoring. a-case-study-in-security-big-data-analysis/d/d-id/1137299?. Accessed 2015-1-10.

4.   Chickowski E (2013) Moving Beyond SIEM For Strong Security Analytics. http://www.darkreading.com/moving-beyond-siem-for-strong-security-analytics/d/d-id. 1141069?. Accessed 2015-1-10.

5.   Marko K (2014) Big Data: Cyber Security's Silver Bullet? Intel Makes the Case. http://www.forbes.com/sites/kurtmarko/2014/11/09/big-data-cyber-security/. Accessed 2015-1-10.

6.   Kezunovic M, Xie L, Grijalva S: The role of big data in improving power system operation and protection. In *Bulk Power System Dynamics and Control - IX Optimization, Security and Control*

of the Emerging Power Grid (IREP), 2013 IREP Symposium. IEEE, Rethymno, Greece; 2013:1-9.Publisher Full Text

7. Software I (2013) Managing big data for smart grids and smart meters. http://www-935.ibm.com/services/multimedia/ Managing_big_data_for_smart_grids.   and_smart_meters.pdf. Accessed 2015-1-10.

8. Modi C, Patel D, Borisaniya B, Patel H, Patel A, Rajarajan M: A survey of intrusion detection techniques in cloud. *J Netw Comput Appl* 2013, 36(1):42-57.

9. Zhou CV, Leckie C, Karunasekera S: A survey of coordinated attacks and collaborative intrusion detection. *Comput Secur* 2010, 29(1):124-140.

10. Laney D (2001) 3d data management: Controlling data volume, velocity and variety. Technical Report 949, META Group (now Gartner).   http://blogs.gartner.com/doug-laney/files/2012/01/ ad949-3D-Data-Management-Controlling-Data-Volume-Velocity-and-Variety.pdf.

11. Zikopoulos P, Parasuraman K, Deutsch T, Giles J, Corrigan D: *Harness the power of big data The IBM big data platform*. McGraw Hill Professional, New York, NY; 2012.

12. Frank J: Artificial intelligence and intrusion detection: current and future directions. In*Proceedings of the 17th national computer security conference. Vol. 10*. Citeseer, Baltimore, MD, USA; 1994:1-12.

13. Information Assurance Solutions Group (2015) Defense in depth. Technical report, National Security Agency. http://www.nsa.gov/ ia/_files/support/defenseindepth.pdf. Accessed 2015-1-10.

14. Denning DE: An intrusion-detection model. *Softw Eng IEEE Trans* 1987, SE-13(2):222-232. doi:10.1109/TSE.1987.232894

15. Sourcefire (2015) Snort, Home Page. http://www.snort.org/. Accessed 2015-1-10.

16. Roesch M: Snort: Lightweight intrusion detection for networks. In *LISA. Vol. 99*. USENIX, Seattle, WA, USA; 1999:229-238.

17. Center for Strategic and International Studies (2013) The economic impact of cybercrime and cyber espionage. Technical report. McAfee http://www.mcafee.com/us/resources/reports/rp-economic-impact-cybercrime.pdf.

18.   Verizon RISK Team (2013) 2013 data breach investigations report. Technical report. Verizon http://www.verizonenterprise.com/resources/reports/rp_data-breach-investigations-report-2013_en_xg.pdf.

19.   Ponemon Institute LLC (2012) 2012 cost of cyber crime study: United states. Technical report. Ponemon Institute http://www.ponemon.org/local/upload/file/2012_US_Cost_of_Cyber_Crime_Study_FINAL6\%20.pdf.

20.   Julisch K, Dacier M: Mining intrusion detection alarms for actionable knowledge. In *Proceedings of the Eighth ACM SIGKDD International Conference on Knowledge Discovery and Data Mining*. ACM, Edmonton, Alberta, Canada; 2002:366-375.

21.   Xu D, Ning P: Correlation analysis of intrusion alerts. *Intrusion Detect Syst* 2008, 38:65-92.

22.   Suthaharan S, Panchagnula T: Relevance feature selection with data cleaning for intrusion detection system. In *Southeastcon, 2012 Proceedings of IEEE*. IEEE, Orlando, FL, USA; 2012:1-6.

23.   Bhatti R, LaSalle R, Bird R, Grance T, Bertino E: Emerging trends around big data analytics and security: Panel. In *Proceedings of the 17th ACM Symposium on Access Control Models and Technologies. SACMAT '12*. ACM, New York, NY, USA; 2012:67-68. doi:10.1145/2295136.2295148. http://doi.acm.org/10.1145/2295136.2295148

24.   Oltsik J (2013) Defining Big Data Security Analytics. Networking Nuggets and Security Snippets (Blog). http://www.networkworld.com/community/blog/defining-big-data-security-analytics. Accessed 2014-5-23.

25.   Sommer R, Paxson V: Outside the closed world: On using machine learning for network intrusion detection. In *Security and Privacy (SP), 2010 IEEE Symposium On*. IEEE, Oakland, CA, USA; 2010:305-316.

26.   Coull SE, Wright CV, Monrose F, Collins MP, Reiter MK: Playing devil's advocate: Inferring sensitive information from anonymized network traces. In *NDSS. Vol. 7*. Internet Society, San Diego, CA, USA; 2007:35-47.

27.   Azad C, Jha VK: Data mining in intrusion detection: a comparative study of methods, types and data sets. *Int J Inf Technol Comput Sci* 2013, 5(8):75-90.

28. McHugh J: Testing intrusion detection systems: a critique of the 1998 and 1999 darpa intrusion detection system evaluations as performed by lincoln laboratory. *ACM Trans Inf Syst Secur* 2000, 3(4):262-294.

29. Mahoney MV, Chan PK: An analysis of the 1999 darpa/lincoln laboratory evaluation data for network anomaly detection. In *Recent advances in intrusion detection*. Springer, Berlin Heidelberg; 2003:220-237.

30. Wu SX, Banzhaf W: The use of computational intelligence in intrusion detection systems: A review. *Appl Soft Comput* 2010, 10(1):1-35.

31. Shiravi A, Shiravi H, Tavallaee M, Ghorbani AA: Toward developing a systematic approach to generate benchmark datasets for intrusion detection. *Comput Secur* 2012, 31(3):357-374. doi:10.1016/j.cose.2011.12.012

32. Fontugne R, Borgnat P, Abry P, Fukuda K: Mawilab: Combining diverse anomaly detectors for automated anomaly labeling and performance benchmarking.[http://doi.acm.org/10.1145/1921168.1921179] *webcite Proceedings of the 6th International COnference. Co-NEXT '10* ACM, New York, NY, USA; 2010, 8-1812. doi:10.1145/1921168.1921179. http://doi.acm.org/10.1145/1921168.1921179

33. United States Marine Academy – West Point (2015) Cyber Research Center – DataSets. http://www.usma.edu/crc/SitePages/DataSets.aspx. Accessed 2015-1-10.

34. Internet Storm Center (2015) Reports – Internet Security | SANS ISC. https://isc.sans.edu/reports.html. Accessed 2015-1-10.

35. Song J, Takakura H, Okabe Y, Eto M, Inoue D, Nakao K: Statistical analysis of honeypot data and building of kyoto 2006+ dataset for nids evaluation. In *Proceedings of the First Workshop on Building Analysis Datasets and Gathering Experience Returns for Security*. ACM, Salzburg, Austria; 2011:29-36.

36. Bass T: Intrusion detection systems and multisensor data fusion. *Commun ACM* 2000, 43(4):99-105.

37. Wang H, Liu X, Lai J, Liang Y: Network security situation awareness based on heterogeneous multi-sensor data fusion and neural network. In *Computer and Computational Sciences, 2007.*

IMSCCS 2007. Second International Multi-Symposiums On. IEEE, Iowa City, IA, USA; 2007:352-359.

38.   Tsang C-H, Kwong S, Wang H: Genetic-fuzzy rule mining approach and evaluation of feature selection techniques for anomaly intrusion detection. *Pattern Recognit* 2007, 40(9):2373-2391.

39.   Chebrolu S, Abraham A, Thomas JP: Feature deduction and ensemble design of intrusion detection systems. *Comput Secur* 2005, 24(4):295-307.

40.   Chen Y, Li Y, Cheng X-Q, Guo L: Survey and taxonomy of feature selection algorithms in intrusion detection system. In *Information Security and Cryptology. Lecture Notes in Computer Science. Vol. 4318.* Edited by Lipmaa H, Yung M, Lin D. Springer, Berlin Heidelberg; 2006:153-167.

41.   Elngar A, Mohamed D, Ghaleb F: A real-time anomaly network intrusion detection system with high accuracy. *Inf Sci Lett Int J* 2013, 2(2):49-56.

42.   The Apache Software Foundation (2015) Welcome to Apache Hadoop!. http://hadoop.apache.org/. Accessed 2015-1-10.

43.   Suthaharan S: Big data classification: problems and challenges in network intrusion prediction with machine learning. In *Big Data Analytics Workshop, in Conjunction with ACM Sigmetrics.* ACM, Pittsburgh, PA, USA; 2013.

44.   Whitworth J, Suthaharan S: Security problems and challenges in a machine learning-based hybrid big data processing network systems. In *ACM Sigmetrics 2013 (Big Data Analytics Workshop).* ACM, Pittsburgh, PA, USA; 2013.

45.   Jeong H, Hyun W, Lim J, You I: Anomaly teletraffic intrusion detection systems on hadoop-based platforms: A survey of some problems and solutions. In *Network-Based Information Systems (NBiS), 2012 15th international conference on.* IEEE, Melbourne, Australia; 2012:766-770.

46.   Lee Y, Lee Y: Toward scalable internet traffic measurement and analysis with hadoop. *ACM SIGCOMM Comput Commun Rev* 2013, 43(1):5-13.

47.   Cheon J, Choe T-Y: Distributed processing of snort alert log using hadoop. *Int J Eng Technol(0975-4024)* 2013, 5(3):2685-2690.

48. VeetiL S, Gao Q: A real-time intrusion detection system by integrating hadoop and naive bayes classification. In *Dalhousie Computer Science In-house Conference (DCSI)*. Dalhousie University, Halifax, Canada; 2013.

49. Bass T: Multisensor data fusion for next generation distributed intrusion detection systems. In *IRIS National Symposium*. IRIS National Symposium, Laurel, MD, USA; 1999.

50. Lan F, Chunlei W, Guoqing M: A framework for network security situation awareness based on knowledge discovery. In *Computer Engineering and Technology (ICCET), 2010 2nd international conference on. Vol. 1*. IEEE, Chengdu, China; 2010:1-226.

51. Mitchell HB: *Data fusion: concepts and ideas*. Springer, New York, NY; 2012.

52. Hall DL, Llinas J: An introduction to multisensor data fusion. *Proc IEEE* 1997, 85(1):6-23.

53. Fessi B, Benabdallah S, Hamdi M, Rekhis S, Boudriga N: Data collection for information security system. In *Engineering Systems Management and Its Applications (ICESMA), 2010 second international conference on*. IEEE, Sharjah, United Arab Emirates; 2010:1-8.

54. Karim Ganame A, Bourgeois J, Bidou R, Spies F: A global security architecture for intrusion detection on computer networks. *Comput Secur* 2008, 27(1):30-47.

55. Bye R, Camtepe SA, Albayrak S: Collaborative intrusion detection framework: characteristics, adversarial opportunities and countermeasures. In *Proceedings of CollSec: Usenix Workshop on Collaborative Methods for security and privacy*. USENIX, Washington, DC, USA; 2010.

56. Bartos K, Rehak M: Self-organized mechanism for distributed setup of multiple heterogeneous intrusion detection systems. In *Self-Adaptive and Self-Organizing Systems Workshops (SASOW), 2012 IEEE sixth international conference on*. IEEE, Lyon, France; 2012:31-38.

57. Cai H, Wu N: Design and implementation of a dids. In *2010 IEEE International Conference on Wireless Communications, Networking and Information Security*. IEEE, Beijing, China; 2010:340-342.

58. Vincent Zhou C, Leckie C, Karunasekera S: Decentralized multi-dimensional alert correlation for collaborative intrusion detection. *J Netw Comput Appl* 2009, 32(5):1106-1123.

59. Metzger S, Hommel W, Reiser H: Integrated security incident management–concepts and real-world experiences. In *IT Security Incident Management and IT Forensics (IMF), 2011 Sixth International Conference On*. IEEE, Stuttgart, Germany; 2011:107-121.

60. Williams A (2007) The Future of SIEM – The market will begin to diverge. http://techbuddha.wordpress.com/2007/01/01/the-future-of-siem-\%E2\%80\%93-the-market-will-begin-to-diverge/.

61. Anuar NB, Papadaki M, Furnell S, Clarke N: An investigation and survey of response options for intrusion response systems (irss). In *Information Security for South Africa (ISSA), 2010*. IEEE, Johannesburg, South Africa; 2010:1-8.

62. Rouse M (2012) security information and event management (SIEM). http://searchsecurity.techtarget.com/definition/security-information-and-event-management-SIEM.

63. Messmer E (2013) Gartner security report: McAfee up, Trend Micro down. http://www.networkworld.com/news/2013/053013-gartner-security-survey-270297.html.

64. Mosaic Security Research Log Management & Security Information and Event Management (SIEM) Software Guide | Mosaic Security Research. http://mosaicsecurity.com/categories/85-log-management-security-information-and-event-management. Accessed 2014-5-23.

65. Aguirre I, Alonso S: Improving the automation of security information management: A collaborative approach. *Secur Privacy IEEE* 2012, 10(1):55-59.

66. Kotenko I, Polubelova O, Saenko I: The ontological approach for siem data repository implementation. In *Green Computing and Communications (GreenCom), 2012 IEEE international conference on*. IEEE, Besancon, France; 2012:761-766.

67. Nicolett M, Kavanagh KM (2011) Critical capabilities for security information and event management technology. Gartner Report.

68. Radack S, Kuhn R (2011) Managing security: the security content automation protocol In: IT Professional. IEEE 9(13):9–11.

69. Ogle D, Kreger H, Salahshour A, Cornpropst J, Labadie E, Chessell M, Horn B, Gerken J, Schoech J, Wamboldt M (2002) Canonical situation data format: the common base event v1.1.1. IBM Corporation. http://xml.coverpages.org/ IBMCommonBaseEventV111.pdf. Accessed 2015-1-10.

70. Distributed Management Task Force Inc (2014) Common Information Model (CIM). http://dmtf.org/standards/cim. Accessed 2014-5-23.

71. Revelytix Inc. (2010) Triple store evaluation analysis report. Technical report, Revelytix. http://www.algebraixdata.com/ wp-content/uploads/2014/02/Revelytix-Triplestore-Evaluation-Analysis-Results.pdf.

72. Kotenko I, Chechulin A: Common framework for attack modeling and security evaluation in siem systems. In *Green Computing and Communications (GreenCom), 2012 IEEE international conference on*. IEEE, Besancon, France; 2012:94-101.

73. Kreutz D, Casimiro A, Pasin M: A trustworthy and resilient event broker for monitoring cloud infrastructures. In *Distributed applications and interoperable systems*. Springer, Berlin Heidelberg; 2012:87-95.

74. Splunk Inc.Operational Intelligence, Log Management, Application Management, Enterprise Security and Compliance | Splunk. http://www.splunk.com/. Accessed 2014-5-23.

75. Li Y, Liu Y, Zhang H: Cross-boundary enterprise security monitoring. In *Computational Problem-Solving (ICCP), 2012 international conference on*. IEEE, Leshan, China; 2012:127-136.

76. Blum D, Schacter P, Maiwald E, Krikken R, Henry T, de Boer M, Chuvakin A (2011) 2012 planning guide: Security and risk management. Technical Report G00224667 Gartner, Inc.

77. Sitaram D, Sharma M, Zain M, Sastry A, Todi R: Intrusion detection system for high volume and high velocity packet streams: A clustering approach. *Int J Innovation Manag Technol* 2013, 4(5):480-485.

78. Kaszuba G (2013) packetloop/packetpig.GitHub.0 https://github. com/packetloop/packetpig.

79.   Yen T-F, Oprea A, Onarlioglu K, Leetham T, Robertson W, Juels A, Kirda E: Beehive: large-scale log analysis for detecting suspicious activity in enterprise networks. In *Proceedings of the 29th Annual Computer Security Applications Conference*. ACM, New Orleans, LA, USA; 2013:199-208.

80.   Myers J, Grimaila MR, Mills RF: Log-based distributed security event detection using simple event correlator. In *System Sciences (HICSS), 2011 44th Hawaii International Conference on*. IEEE, Kauai, HI, USA; 2011:1-7. Publisher Full Text

81.   Langner R: Stuxnet: Dissecting a cyberwarfare weapon. *Secur Privacy IEEE* 2011, 9(3):49-51.

82.   Valdes A, Cheung S: Intrusion monitoring in process control systems. In *System Sciences, 2009. HICSS'09. 42nd Hawaii international conference on*. IEEE, Waikoloa, Big Island, HI, USA; 2009:1-7.

83.   SRI International (2014) Detection and Analysis of Threats to the Energy Sector (DATES). http://www.csl.sri.com/projects/dates/. Accessed 2014-5-23.

84.   Valdes A (2010) Detection and analysis of threats to the energy sector: Dates. Technical report, SRI International.

85.   XU X-b, YANG Z-q, XIU J-p, LIU C: A big data acquisition engine based on rule engine. *J China Universities Posts Telecommunications* 2013, 20:45-49.

86.   Ray PD, Reed C, Gray J, Agarwal A, Seth S (2012) Improving roi on big data through formal security and efficiency risk management for interoperating ot and it systems In: Grid-Interop Forum 2012, Irving, Texas, USA.

87.   Gabriel R, Hoppe T, Pastwa A, Sowa S: Analyzing malware log data to support security information and event management: Some research results. In *Advances in databases, knowledge, and data applications, 2009. DBKDA'09. First international conference on*. IEEE, Cancun, Mexico; 2009:108-113.

88.   Hunt R, Slay J: The design of real-time adaptive forensically sound secure critical infrastructure.In *Network and System Security (NSS), 2010 4th International conference on*. IEEE, Melbourne, Australia; 2010:328-333.

89. Sundaramurthy SC, Bhatt S, Eisenbarth MR: Examining intrusion prevention system events from worldwide networks. In *Proceedings of the 2012 ACM workshop on building analysis datasets and gathering experience returns for security*. ACM, Raleigh, NC, USA; 2012:5-12.

90. Roschke S, Cheng F, Meinel C: A flexible and efficient alert correlation platform for distributed ids. In *Network and System Security (NSS), 2010 4th international conference on*. IEEE, Melbourne, Australia; 2010:24-31.

Mode I Critical Fracture Energy of Adhesively Bonded Joints between Glass Fibers Reinforced Thermoplastics

Siripong Mahaphasukwat[1], Kazumasa Shimamoto[1],
Shota Hayashida[1], Yu Sekiguchi[2], and Chiaki Sato[2]

[1]Graduate school, Tokyo Institute of Technology, 4259 Nagatsuta, Midori-ku, Yokohama 226-8503, Japan
[2]Precision and Intelligence Laboratory, Tokyo Institute of Technology, 4259 Nagatsuta, Midori-ku, Yokohama 226-8503, Japan

ABSTRACT

Critical fracture energies of adhesively bonded joints under mode I constant separation were experimentally investigated. Double cantilever beam (DCB) specimens comprising polyamide 6 (PA6) based fiber reinforced thermoplastics (GFRTP) were utilized for the experiments. The adherends of the joints were bonded with three different types of adhesives such as polyurethane and acrylates. A surface treatment method with a primer was applied to pre-bonded surface, matching with the different adhesives, which results in five combinations.

Strongest combination, Plexus Primer PC120 and Plexus AO420, exhibited 2.95 kJ/m^2 in mode I critical fracture energy, which is much higher than those of ordinary epoxy adhesive and similar to those of rubber-modified very-ductile epoxy adhesives. Therefore, it is confirmed that adhesive bonding can be applied to join PA6 based GFRTP even for structural use, although the material is thought too difficult to bond adhesively.

BACKGROUND

Adhesive bonding technology and applications for composite materials are of particular importance to many industries because of their ability to support and improve the features of future's products such as light-weight transportations. Let us take the automotive industry as an example; the steel car structure is mainly used in present day automotive industries. Substituting steel with aluminum alloy or composite materials wherever possible can provide many benefits to a car performance, such as higher fuel efficiency by weight reduction, stiffer chassis and manageable weight distribution for better handling, design variety, etc. [1],[2]. Composite materials such as glass fiber or carbon fiber reinforced plastics (GFRP or CFRP) are the most promising in terms of reducing the weight of a car body in white.

The application of adhesive bonding is also very beneficial because it makes the bonding between different materials possible and it also provides more uniform stress distribution in the joint area over conventional mechanical fasteners that expose the material to concentrated stress [3]. For composite materials, reducing stress concentration in joints is crucial, and dissimilar materials joining with metals is also ineluctable to fabricate real car structures. Thus, adhesive bonding is very promising as one of joining methods for the future's car structures consisting of composite materials

The matrix resin for fiber reinforced plastics (FRP) is gradually changing from thermosetting type such as epoxy resins to thermoplastic type such as polyamide resins because the forming time of thermoplastic composites is much shorter than that of thermoplastic composites, leading to shorter cycle times in assembly lines and increased efficiency in production, which is indispensable for the automotive industry.

There are many bonding methods that can be applied to the thermoplastic composites, such as welding. Even for dissimilar materials with thermoplastic composites, welding can be applied as thermo-melted fusion bonding methods [4],[5]. On the other hand, even though the use of adhesives is still a promising joining method for thermoplastic composites [6], it is thought that those materials are hard to bond because of their low surface energies. To identify the most reliable bonding method that can be applied in the making of structures, consideration in terms of cost, time, and efficiency, should be investigated.

However, research on adhesively bonded joints of thermoplastic composites is incipient still now. Therefore, this research focuses on the strength of joints between glass fiber reinforced thermoplastics (GFRTP) as the adherends, bonded with three different types of adhesive. Further study and exploration on the use of different surface pretreatments, such as primer pretreatment matching to various kinds of adhesives available in market, have been carried out.

Furthermore, methods to measure the strength of the bonded joints are also very important. In the past, designs for engineering structures have been dominated by using approaches based on mechanics of materials, in which allowable stress or strains are applied as the strength criteria. However, such approaches have a difficulty because stress singularity may occur near the edges of adhesive layer and that leads to the dependency of predicted strength on the mesh size for finite element methods. Recent approaches on the design for strength of structures, fracture mechanics offer several criteria for evaluating the strength of structures including flaws [7] or adhesively bonded joints [8]-[14].

To test for the strength of adhesively bonded joints in this research, double cantilever beam (DCB) specimens were prepared and used in the experiments. Based on linear elastic fracture mechanic (LEFM), the energy release rate approach was applied to obtain the critical fracture energies of the joints. Fracture in adhesive layer may occur in three different loading modes: mode I (opening), mode II (forward shear), and mode III (tearing). However, this research will focus on the mode I loading condition.

METHODS

Materials

Double cantilever beam specimens were prepared for the tests. The specimen had two adherends made of a glass fiber reinforced thermoplastic (TEPEX® Dynalite 102-RG600(6)/47%-3.0 mm, Bond-Laminates GmbH, Germany) including polyamide 6 (PA6) matrix resin. Table 1 shows the mechanical properties of the GFRTP. Three types of adhesives, one polyurethane and two acrylates, were selected for the research, considering the quick curing capability at ambient temperature necessary for the automotive industry. A surface primer was also used combined with the acrylate adhesives, so that the combination number of adhesives and primer was five. The primer and the adhesives used for the experiments, and their designation are as follows;

Table 1: GFRTP properties (provided by manufacturer)

Material	Fibre	Polymer	Tensile Strength (MPa)	Tensile Modulus (GPa)	Flexural Strength (MPa)	Flexural Modulus (GPa)	Processing temperature (°C)	Main application use
TEPEX® dynalite 102	Roving Glass	PA6 (Polyamide)	405	22	620	19	240	automotive, protection, consumer, sports, miscellaneous

Mahaphasukwat et al.

Mahaphasukwat et al. Applied Adhesion Science 2015 3:4, doi:10.1186/s40563-015-0036-2

Primer P (Plexus Primer/Conditioner PC120, Illinois Tool Works Inc., USA), which was designed to improve long term durability for adhesively bonded joins with acrylate adhesive when used for aluminum or stainless steel assemblies [15].

Adhesive A (Sikaflex-252, Sika AG, Switzerland) is a 1-component, moisture cured, polyurethane adhesive [16].

Adhesive B (Plexus MA300, Illinois Tool Works Inc., USA) is a two-part methacrylate adhesive designed for structural bonding with high strength and stiffness as well as the ability to bond a wide range of materials [17].

Adhesive C (Plexus AO420, Illinois Tool Works Inc., USA) is also a two-part methacrylate adhesive designed for structural bonding. It provides a unique combination of high strength, good fatigue endurance, high impact resistance, and toughness [18].

In this paper, the combinations of adhesives and primer are abbreviated and denoted as follows:

- Adhesive A without Primer: A
- Adhesive B without Primer: B
- Adhesive B with Primer P: BP
- Adhesive C without Primer: C
- Adhesive C with Primer P: CP

Specimen preparation

To make DCB specimens shown in Figure 1, two large GFRTP plates, whose size is 180×200 mm^2, were bonded and cut into the specimens. To create a new fresh surface on the CFRP plate, the surfaces were sandblasted and cleaned with acetone to remove all the contaminants that can affect the bonding strength. Primer P was applied to the specimens BP and CP. Since only a thin film of Primer P is necessary and required, the primer was applied and wiped to control the thickness as thin as possible.

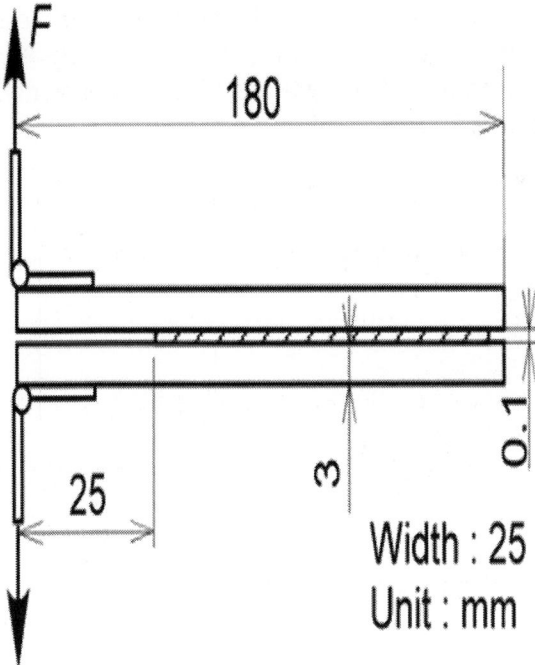

Figure 1: Specimen configuration.

The bonding procedure is as follows:

- Releasing film (Teflon) of 25×200 mm² was placed on one side of a GFRTP plate to ensure pre-crack of 25 mm-length.

- A narrow releasing film (Teflon) of approximately 10×200 mm² was placed on the other end as a shim to ensure an adhesive layer thickness of approximately 0.1 mm.

- The specific type of adhesive was applied on the surface and spread evenly throughout.

- The other GFRTP plate was placed on the top.

- The plates were inserted into a silicon dam surrounding them in order to achieve proper alignment and were transferred to a hydraulic press, as shown in Figure 2.

- Sufficient pressure, approximately 0.53 MPa (15 MPa-gage pressure), was applied to the area (25×200 mm²) to achieve good adhesive distribution over the surfaces of the plates and kept for the required time to cure each adhesive.

- The bonded plates were released and cut it into specimen size of 25 mm in width.
- Two piano hinges were installed on the 25 mm pre-cracked side to an obtained specimen of the final shape as shown in Figure 1.

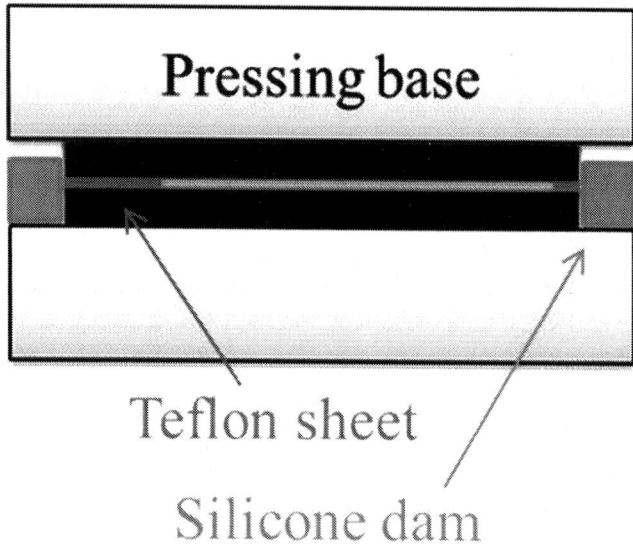

Figure 2: Schematic illustration for specimen fabrication.

Experimental procedure

For mode I tests of the DCB specimens, a mechanical material testing machine (AGS-500A, Shimadzu Co., Ltd., Japan) was utilized. The specimen was fixed with mechanical grips, as shown in Figure 3, and loaded in the tensile direction with a constant displacement rate of 5 mm/min. The load and displacement were simultaneously recorded with a frequency of 1 Hz. For each type of specimen, three tests were carried out to obtain average value and standard deviation of critical fracture energy.

Figure 3: Mode I test of DCB specimen consisting of GFRTP.

RESULTS AND DISCUSSION

Figure 4 shows fractured surfaces of specimens A, B, BP, C and CP. For specimens A, B and BP, adhesive fracture between adherends and adhesives was mostly observed. In contrast, mix of adhesive and cohesive fracture was observed on the fracture surfaces of specimens C and CP. Polyamide 6, which is the matrix resin of the adherends, has relatively small amount of functional groups that contribute to adhesion, comparing from other thermoset matrix resins such as epoxy and polyester. Therefore, the interfacial strengths between adhesive and the adherends are low, and adhesive fracture occurs easily. Primer P can chemically activate the surface of the adherends and increases the interfacial strength to adhesive. The stronger interface leads to the cohesive fracture for specimens CP.

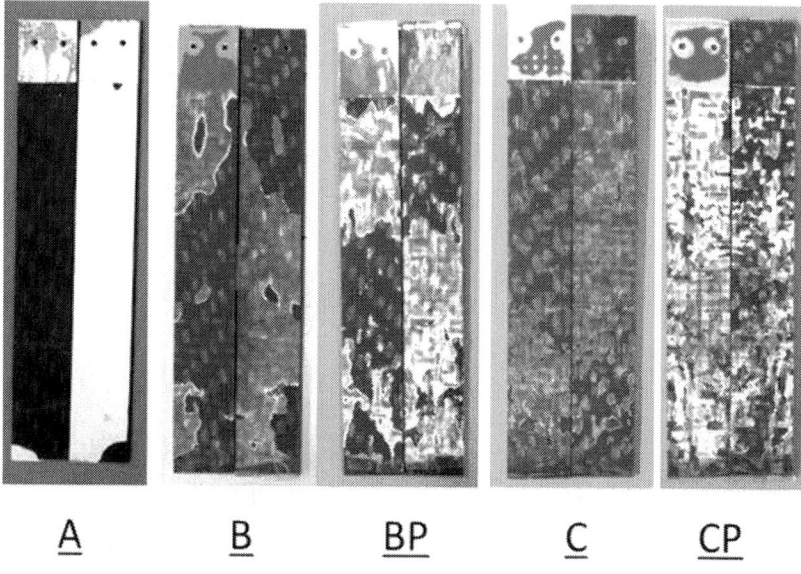

A   B   BP   C   CP

Figure 4: Fracture surfaces of different types of specimens.

Load–displacement curves of specimens A, B, BP, C and CP are shown in Figures 5, 6, 7, 8, 9. These figures include all the data for each experiment and suffixes 1, 2 and 3 indicate the specimen number. Specimens A and B exhibited similar load–displacement curves, as shown in Figures 5 and 6. The load–displacement curves of specimens A were quite smooth because crack propagation occurred continuously, and it resulted in small deviation in load and high repeatability. In contrast, for specimens B, crack propagated intermittently, which leaded to large deviation and low repeatability. Specimen C had higher maximum load and displacement, as shown in Figure 8, so that adhesive C is stronger and more ductile than adhesives A and B. Comparing Figures 6 and 8 with Figures 7 and 9, the maximum loads and displacements largely increased using primer P with adhesives B and C.

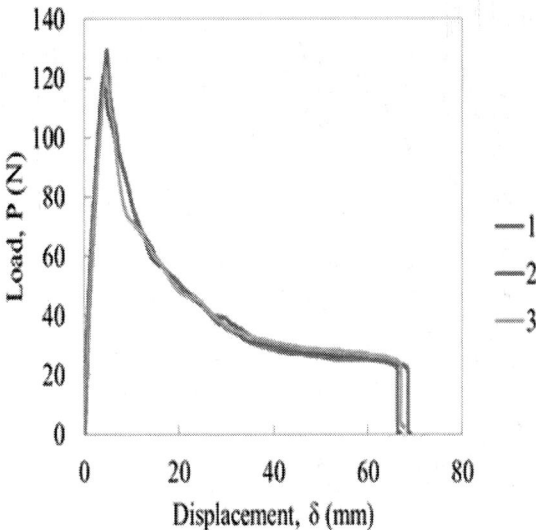

Figure 5: Load–displacement curves from specimens bonded with adhesive A.

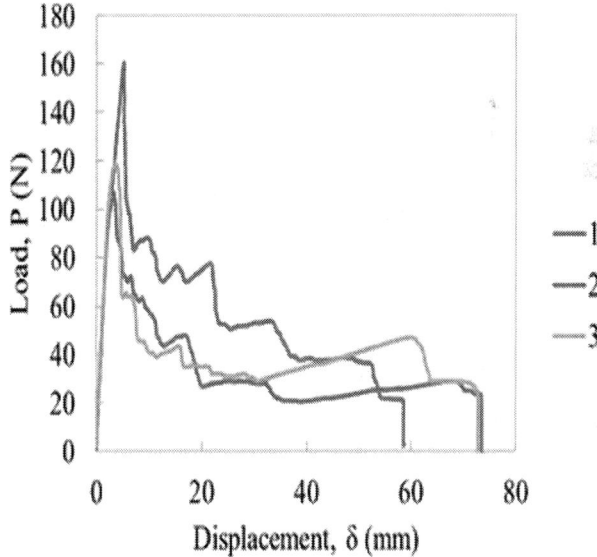

Figure 6: Load–displacement curves from specimens bonded with adhesive B.

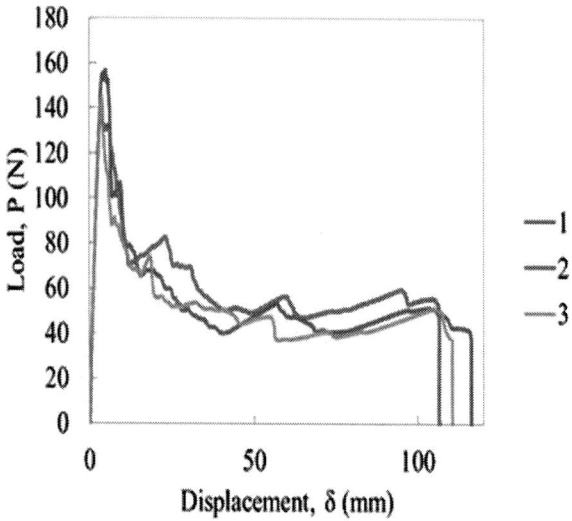

Figure 7: Load–displacement curves from specimens bonded with adhesive BP.

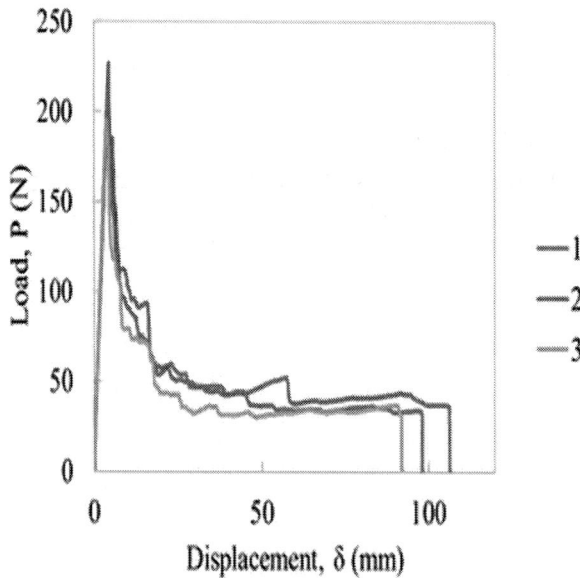

Figure 8: Load–displacement curves from specimens bonded with adhesive C.

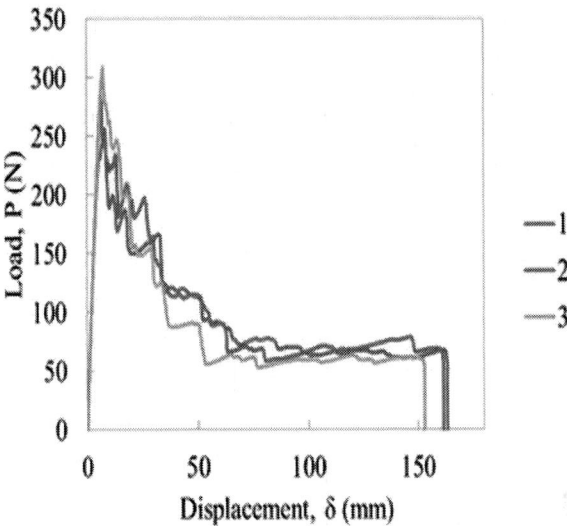

Figure 9: Load–displacement curves from specimens bonded with adhesive CP.

To calculate the critical fracture energy of the specimens, an approach based on LEFM was applied. Energy release rate, g, which is the amount of energy per unit crack area created by a growing crack can be shown as for systems which energy dissipation is limited to the crack tip region. Here, W is the external work, U is the stored elastic energy, and A is the crack area. The crack will propagate when this applied energy release rate reaches the critical value, g_C, related to the fracture toughness in mode I, g_I.

$$g = \frac{\partial(W-U)}{\partial A}$$

(1)

For a bonded configuration in which the load and deflection are linearly related, the value of mode I energy release rate g_I is given by, according to LEFM:

$$gI = \frac{P^2}{2B}\frac{dC}{da}$$

(2)

where P is the load, B is the width of DCB specimen, a is the crack length, and C is the compliance, given by

$$C = \delta/P \qquad (3)$$

where   is the displacement corresponding to a load P. From simple beam theory, the value of the compliance C is given by

$$C = \frac{\delta}{P} = \frac{2a^3}{3EI} \qquad (4)$$

where E is the young's modulus of adherend and I is the second moment of area, $I = Bh^3/12$, where h is the thickness of adherend. Then the energy release rate in mode I becomes

$$g_I = \frac{P^2 a^2}{BEI} = \frac{12P^2 a^2}{B^2 h^3 E} \qquad (5)$$

In the experiments, since relatively soft adhesives were utilized, the crack tips were difficult to identify visually, the process zones were quite large. Thus, it was difficult to calculate g_I from visually observed crack length. Chaves et al. proposed a crack equivalent method by energy release rate can be determined without crack length even for mixed mode conditions [19]. Based on the theory proposed by Chaves and the simple beam theory, a following method that is simpler and can be used only for mode I loading was derived and applied to the test results.

To eliminate the need of monitoring the change in crack length for the ease of result analysis, substituting a by applying simple beam theory (modification of Eq. (4)), where a is given by

$$a = \sqrt[3]{\frac{3EI\delta}{2P}} \qquad (6)$$

By substituting Eq. 6 into Eq. 5, mode I energy release rate can be expressed by

$$g_I = \frac{P^2}{B^3 \sqrt{EI}} \left(\frac{3\delta}{2P}\right)^{2/3} \qquad (7)$$

which the crack length, a, is no longer required for energy release rate calculation. Equation 6 is applicable only to DCB specimens having thin adherends whose deformation can be explained by the simple beam theory. That is the limitation of the proposed method.

Mode I critical fracture energy g_{IC} vs. displacement curves of the specimens were then obtained based on Eq. (7), as shown in Figures 10, 11, 12, 13, 14 accordingly. This varied curves were then averaged by using area under the curve divided by displacement and the results are denoted by $g_{IC,avg}$. The results of three experiments for each adhesive are shown in Table 2. Three results of $g_{IC,avg}$ were then again averaged ($g_{IC,avg}$ average) in each type of specimens and are shown in Table 2 and Figure 15. Comparing all the results of critical fracture energy, adhesive C (Plexus AO420) had the highest value, adhesive A (Sikaflex-252) was in the second position and adhesive B (Plexus MA300) in the third, although the difference between those of adhesives A and B was not significant. The average critical fracture energy of adhesive C was 918 J/m², which is not smaller than those of ordinary epoxy adhesives, and the values of adhesive B, which was the smallest, was 403 J/m² that is quite similar to the typical critical fracture energy of brittle epoxy adhesive. Thus, the used adhesives, polyurethane and acrylate, are not inferior to epoxy adhesives in terms of critical fracture energy for joining PA6 based GFRTP. They may be utilized for structural purposes instead of epoxy adhesives.

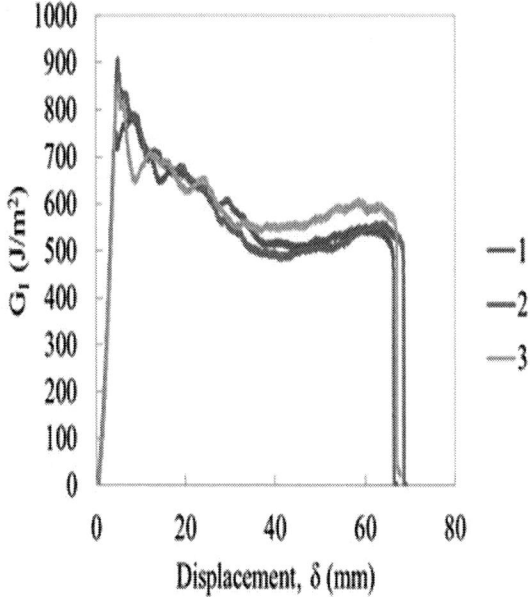

Figure 10: Critical fracture energies with respect to displacement from specimens bonded with adhesive A.

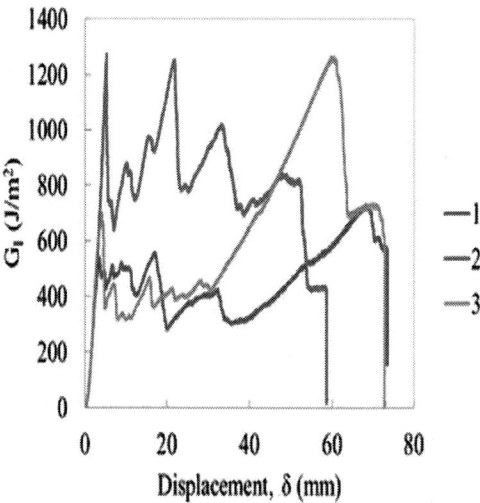

Figure 11: Critical fracture energies with respect to displacement from specimens bonded with adhesive B.

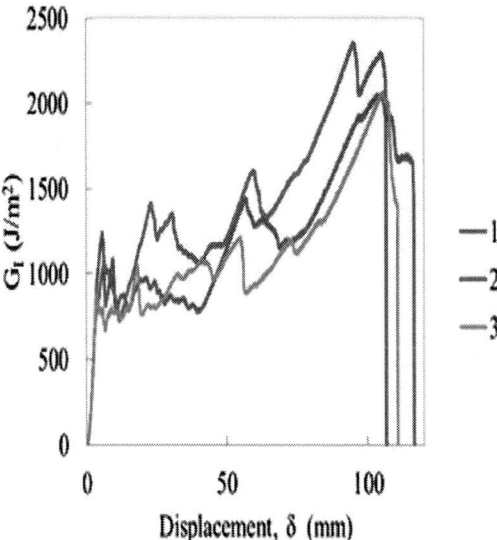

Figure 12: Critical fracture energies with respect to displacement from specimens bonded with adhesive BP.

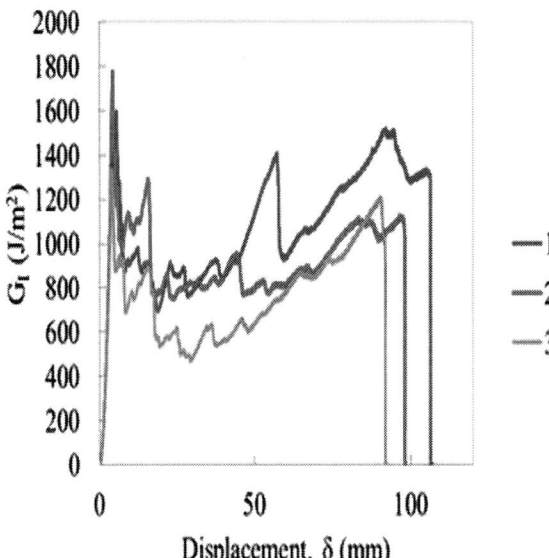

Figure 13: Critical fracture energies with respect to displacement from specimens bonded with adhesive C.

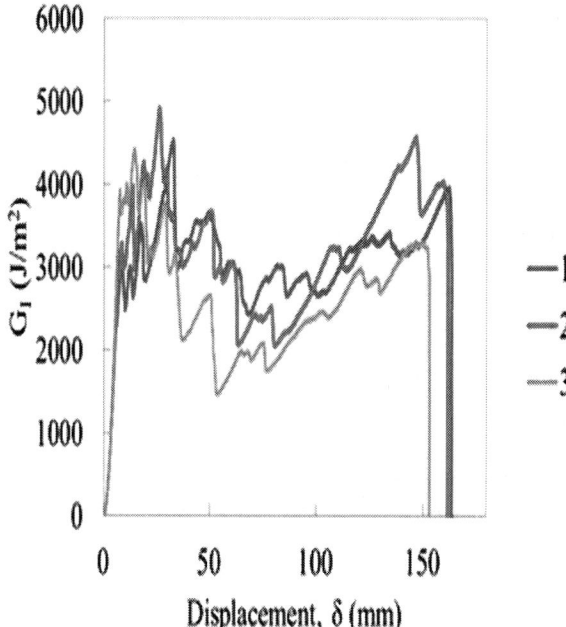

Figure 14: Critical fracture energies with respect to displacement from specimens bonded with adhesive CP.

Table 2: Experiment results

Exp. No.	$g_{IC,avg}$ (kJ/m^2)				
	A	**B**	**B P**	**C**	**C P**
1	0.574	0.348	1.24	1.08	3.08
2	0.568	0.464	1.43	0.918	3.18
3	0.589	0.396	1.14	0.760	2.58
Average	0.577	0.403	1.27	0.918	2.95
S.D.	0.0108	0.0582	0.150	0.158	0.321

Mahaphasukwat et al.

Mahaphasukwat et al. Applied Adhesion Science 2015 3:4, doi:10.1186/s40563-015-0036-2

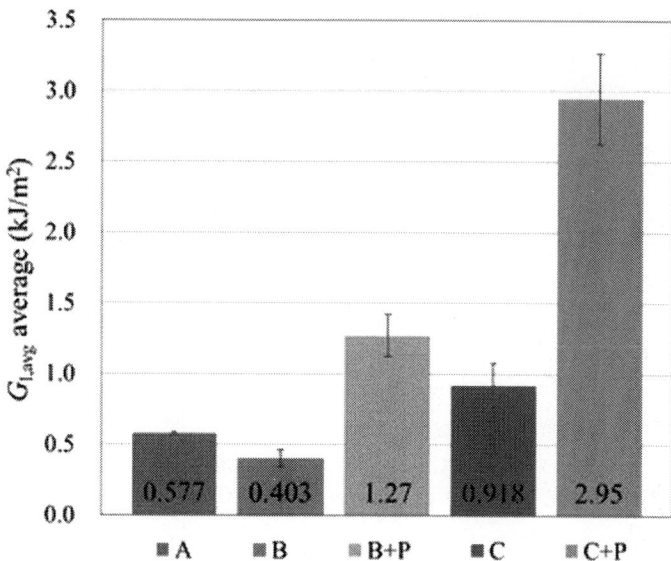

Figure 15: Comparison of overall-average mode I fracture critical energies.

The effect of primer P was drastic because the use increased the critical fracture energies of adhesive B and C approximately three times. The results show that the combination of adhesive C (Plexus AO420) and primer P (Plexus Primer PC120) exhibited the strongest value of 2.95 kJ/m² that is much higher than those of ordinary epoxy adhesives and not inferior to the critical fracture energy of the most ductile epoxy adhesives such as CTBN modified epoxy adhesives.

Adhesive A (Sikaflex-252), which is polyurethane, was not combined with any primer in this research. The reason is only due to the situation that the authors did not have any primer appropriate for polyurethane adhesives. Polyurethane adhesives are usually utilized with surface treatment methods such as flame treatments and surface primers when applied to thermoplastics because the materials have low surface energy and are difficult to bond. The possibility that adhesive A exhibits higher strength with surface treatments cannot be denied. Thus, the results of this research do not imply the superiority of acrylate adhesives to polyurethane adhesives, but demonstrate the applicability of those types of adhesives to structural use.

CONCLUSIONS

In this research, the strength of adhesively bonded joints between PA6 based CFRTP adherends was experimentally investigated in terms of mode I critical fracture energy. The adherends were bonded with three different types of adhesives: adhesive A (Sikaflex-252), adhesive B (Plexus MA300) and adhesive C (Plexus AO420). A surface pretreatment with primer P (Plexus Primer PC120) matching to the use of three types of adhesives were carried out and studied for its applicability. When it comes to strength measuring in adhesive bonded joints, rather than considering in stress-base method, an approach to the design for strength of a structure, based on linear elastic fracture mechanic (LEFM) were applied. Mode I DCB tests were conducted to study and confirm joint strength from the various bonding methods. From this research, the following conclusions can be obtained.

- PA 6 based thermoplastic composites can be bonded by adhesive. The strength is not weak even if surface treatment is not applied and very high when proper surface treatment is applied and it matches to the type of adhesive.

- Acrylate adhesives B (Plexus MA300) and C (Plexus AO420) have enough strength compared with ordinary epoxy adhesives. The critical fracture energies without primer treatment are 403 J/m^2 for adhesives B and 918 J/m^2 for adhesives C.

- Polyurethane adhesive A (Sikaflex-252) has a critical fracture energy of 577 J/m^2, which is higher than that of adhesives B.

- When primer P (Plexus Primer PC120) is used, the critical fracture energy of DCB specimens increases much. For instance, adhesives B with primer P and adhesives C with primer P exhibited 1.27 kJ/m^2 and 2.95 kJ/m^2 in critical fracture energy, respectively. The values are approximately three times to those without primer treatment.

- The critical fracture energy with adhesive C and primer P, i.e. 2.95 kJ/m^2 is not inferior to the maximum values obtained from sophisticated ductile epoxy adhesives modified with rubber particles.

AUTHORS' CONTRIBUTIONS

SM and KS made the specimen and carried out the experiments, SH prepared the experiment setup, YS and CS help for SM and KS to derive the equations and to write the paper.

ACKNOWLEDGEMENTS

Sika Japan Ltd. and ITW Performance Polymers & Fluids Japan are greatly acknowledged for providing us adhesives.

REFERENCES

1. Beardmore P, Johnson C (1986) The Potential for Composites in Structural Automotive. Compos. Sci. Technol 26:251-281

2. Feraboli P, Masini A (2004) Development of carbon/epoxy structural components. Composites: Part B 35:323-330

3. Barnes T, Pashby I (2000) Joining techniques for aluminium spaceframes used in automobiles. J Mater Process Technol 99:72-79

4. Yousefpour A, Hojjati M, Jean-Pierre I (2004) Fusion Bonding/ Welding of Thermoplastic Composites. J. Thermoplast. Compos. Mater. 17:303-341

5. Ageorges C, Ye L, Hou M (2001) Advances in fusion bonding techniques for joining thermoplastic matrix. Composites: Part A 32:839-857

6. Molitor P, Barron V, Young T (2005) Surface treatment of titanium for adhesive bonding to. Int. J. Adhes. Adhes. 21:129-136

7. Anderson T (2005) Fracture Mechanics. CRC Press, Florida.

8. Dillard DA (2005) Fracture mechanics of adhesive bonds. in Adhesive bonding, ed. R. D. Adams, 190–208. Woodhead Publishing, Cambridge.

9. Blackman BRK, da Silva LFM, Ochsner A, Adams RD (2011) Fracture Tests. In: Handbook of Adhesion technology. Springer-Verlag, Heidelberg. pp 474-501

10. Blackman BRK (2012) Quasi-Static Fracture Tests: Double Cantilever Beam and Tapered Double Cantilever Beam Testing. In: Silva LFM, Dillard D, Blackman B, Adams R (eds) Testing Adhesive Joints, Wiley-VCH Verlag & Co, Weinheim. pp 170-174

11. Blackman BRK, Dear J, Kinloch A, Osiyemi S (1991) The calculation of adhesive fracture energies from double-cantilever beam test specimens. J. Mater. Sci. Lett. 10:253-256

12. Blackman BRK, Kinloch A, Paraschi M, Teo W (2003) Measuring the mode I adhesive fracture energy of structural adhesive joints: the results of an international round-robin. Int. J. Adhes. Adhes. 23:293-305

13. Hashemi S, Kinloch A, Williams J (1990) The analysis of interlaminar fracture in uniaxial fibre-polymer composites. Proceeding of the Royal Society A 427:173-199

14. Williams J (1988) On the calculation of energy release rates for cracked laminates. Int. J. Fract. 36:101-119

15. PC-120 Technical Data Sheet. ITW Plexus, Massachusetts.

16. Sikaflex-252 Elastic Adhesive Technical Data Sheet. Sika Corporation. Michigan.

17. MA300 Technical Data Sheet. ITW Plexus. Massachusetts.

18. AO420 Technical Data Sheet. ITW Plexus. Massachusetts.

19. Chaves FJP, de Moura MFSF, da Silva LFM, Dillard DA (2013) Numerical validation of a crack equivalent method for mixed-mode I + II fracture characterization of bonded joints. Eng. Fract. Mech. 107:38-47

Direct Patterning of Gold Nanoparticles Using Flexographic Printing for Biosensing Applications

Jamie Benson[1], Chung Man Fung[1], Jonathan Stephen Lloyd[1], Davide Deganello[2], Nathan Andrew Smith[3,] and Kar Seng Teng[1]

[1]Multidisciplinary Nanotechnology Centre, College of Engineering, Swansea University, Singleton Park, Swansea SA2 8PP, UK

[2]Welsh Centre for Printing and Coating, College of Engineering, Swansea University, Singleton Park, Swansea SA2 8PP, UK

[3]College of Science, Department of Physics, Swansea University, Singleton Park, Swansea SA2 8PP, UK

ABSTRACT

In this paper, we have presented the use of flexographic printing techniques in the selective patterning of gold nanoparticles (AuNPs) onto a substrate. Highly uniform coverage of AuNPs was selectively patterned on the substrate surface, which was subsequently used in the development of a glucose sensor. These AuNPs provide a biocompatible site for the attachment of enzymes and offer high sensitivity in the detection of glucose due to their large surface to volume ratio. The average size of the printed AuNPs is less than 60 nm. Glucose sensing tests were performed using printed carbon-AuNP electrodes functionalized with glucose oxidase (GOx). The results showed a high sensitivity of 5.52 μA mM^{-1} cm^{-2} with a detection limit of 26 μM. We have demonstrated the fabrication of AuNP-based biosensors using flexographic printing, which is ideal for low-cost, high-volume production of the devices.

BACKGROUND

Noble metal nanoparticles (NPs) have unique electrical, optical, thermal and catalytic properties, which find applications in non-toxic drug delivery systems [1], molecular biosensing [2], cellular imaging [3] and protein detection [4] among others. They have many advantages over other nanomaterials, as they are more stable and conductive [5]. These noble metal NPs can range in sizes from 1 to 100 nm and can be differently shaped [6]. They also have the added advantage of being biocompatible [7]. These types of NPs have been commonly used in molecular diagnostics as they offer excellent sensitivity due to their very large surface to volume ratio and ease of functionalisation for the detection of specific analytes in biological solutions [7]. Among the most extensively used are gold nanoparticles (AuNPs). They have been used in many nanotechnology-based biosensors due to the biocompatibility of the material in addition to its many unique optical and electrical properties. An example of AuNP-based biosensors can be seen in work carried out by Feng et al. where an electrochemical deposition method was used to deposit an AuNP-chitosan film which was subsequently used for the determination of glucose [8]. Another example is work performed by Jena et al. where they introduced AuNPs to a 3D silicate network using a sol-gel process [9]. This was utilised as an enzyme-free biosensor for glucose detection. Their sensor showed high sensitivities of 0.179 nA nM^{-1} cm^{-2} and showed great stability and reproducibility. AuNPs have also been used extensively in enzymatic biosensors. An example of this can be seen in the work of Zhang et al., where they fabricated a glucose sensor using dithiol immersion of a gold electrode followed by immersion in cystamine and AuNPs with subsequent attachment of glucose oxidase (GOx) [10],[11]. Sensors developed in this way showed good sensitivities of 8.8 μA mM^{-1} cm^{-2}[10] and 8.3 μA mM^{-1} cm^{-2}[11]. These sensors highlight the use of AuNPs in highly sensitive biosensors for the detection of not just glucose but a multitude of other biomarkers. However, current methods in fabricating AuNP electrodes are very laborious and time-consuming and are therefore not suitable for scaled-up production of biosensors. This highlights a significant market gap for methods that lend themselves to mass production of AuNP electrodes and thus rendering the device commercially viable. Flexographic printing techniques could overcome these restrictions and help bring highly sensitive AuNP-based sensors to the mass market at low cost in comparison to other techniques. The ability for patients to test themselves for a range of conditions within the comfort of their own home is highly attractive. The point-of-care (PoC) diagnostic device market is poised to reach $27.5 billion by 2018, with a wide range of PoC technologies covering many different diseases and conditions [12]. In order to ensure commercial viability of these technologies, there is a requirement for low cost, high yield fabrication of such devices. The use of printing techniques is an obvious step towards mass production of devices at a relatively low cost when compared to the use of semiconductor cleanroom techniques, which involve multiple processing steps using complex and expensive facilities. Various printing techniques, such as screen printing [13] and inkjet printing [14], have been utilised for biosensor fabrication. Herein, we report the novel use of flexographic printing techniques in the fabrication of AuNP-based devices, such as an electrochemical biosensor. The ability to incorporate many printing rolls allows the printing of multiple layers consecutively via a straightforward printing process. Furthermore, it does not suffer from the 'coffee-ring' effect or clogging of printing heads as observed in the inkjet printing technique [15]. The flexographic printing technique offers a simple and rapid fabrication process for AuNP-based devices.

In this work, carbon electrodes and AuNPs were printed onto a polyimide substrate through the use of flexographic printing. Here, an AuNP ink was developed for the printing technique. Glucose sensors were fabricated, as an exemplar biosensor, to demonstrate the viability of using the flexographic printing technique in the production of AuNP-based electrochemical biosensors. Glucose sensing was chosen due to its popularity and large-scale clinical relevance. The use of GOx for the enzymatic detection of glucose has been utilised for electrochemical glucose sensing for many years, on various electrode constructions, including those incorporating AuNPs. This is due to its well-established selectivity, reliability and relatively low cost [16]-[19]. This printed device could also be used for a multitude of other enzymatic biosensing applications through functionalisation with other enzymes as well as the wealth of other uses for AuNPs discussed above. Sensors fabricated in this work have shown high sensitivity and also displayed a low limit of detection (LoD). GOx immobilised onto the printed AuNPs at the carbon electrodes has shown very good electron transport properties indicated by the fast response time of the sensor to the presence of even small concentrations of glucose.

METHODS

Materials

Polyimide was obtained from Katco (Katco, Milton Keynes, UK) and was cleaned via ultrasonication with acetone prior to use. GOx (259 U mg^{-1}) was purchased from BBI Solutions (BBI Solutions, Cardiff, UK) and was used as received. Glucose was obtained from Fisher Scientific (Fisher Scientific, Loughborough, UK) and made into various concentrations in deionised (DI) water at least 24 h prior to sensor testing to allow for mutarotation. $HAuCl_4 \cdot xH_2O$, polyvinylpyrrolidone (PVP), $NaBH_4$, cysteamine and glutaraldehyde were purchased from Sigma-Aldrich (Sigma-Aldrich, Dorset, UK). Carbon flexographic ink was purchased from Gwent Group (Gwent Group Limited, Pontypool, UK) and used as received. Paraffin wax and Phosphate buffered saline (PBS) solution with pH 7.4 were purchased from Fisher Scientific (Loughborough, UK).

Fabrication of AuNP Ink

AuNPs were synthesised through the chemical reduction of $HAuCl_4$ via $NaBH_4$ addition. A volume 30 ml of DI water was stirred rapidly in a conical flask using a stirrer bar. Next, 0.2 g of HAuCl$_4$ was added to the conical flask followed immediately by 0.15 g of PVP. The PVP acts as a capping agent and prevents the agglomeration of nanoparticles. The solution was allowed to stir for 20 min. A 1-ml aliquot of NaBH$_4$ solution was prepared containing 0.05 g of NaBH$_4$. A few drops of 1 M NaOH were added to the aliquot, and it was placed in the freezer until it reached <5°C. The NaBH$_4$ solution was then added to the conical flask rapidly in two 500-µl aliquots. The nanoparticle reduction was indicated by a solution colour change from bright yellow to dark purple, and, as expected, hydrogen evolution was observed. This solution was allowed to stir for 4 h. After 4 h, the solution was poured into a centrifuge tube and centrifuged at 4,000 rpm for 30 min. After centrifugation, a large pellet of AuNPs was present at the bottom of the centrifuge tube. The supernatant was removed carefully to ensure the pellet was not disturbed. The pellet was then re-dispersed in 70% isopropyl alcohol (IPA) (14 ml) and 30% DI water (6 ml) via ultrasonication. The AuNP ink was then ready for flexographic printing. This is a simple method that only involves a small number of steps. To the knowledge of the authors, AuNP ink has not previously been fabricated through this specific method.

Electrode Preparation

Flexographic printing was performed using an IGT F1 flexographic test printer (IGT Testing Systems, Amsterdam, The Netherlands). A schematic of the system is shown in Figure 1. The working electrode was prepared by first printing carbon ink onto a sheet of polyimide using the flexographic printing technique. After printing carbon onto the polyimide substrate, the samples were annealed at 150°C in an oven for 10 min. The polyimide, with printed carbon, was then cut into 1 cm × 2 cm sections for use as an electrode. The printed carbon sections were then placed back onto the flexographic printer for AuNP printing using the ink described previously. The printing parameters for flexographic printing of our ink were optimised as follows; printing force of 125 N, anilox force of 125 N, and speed of 0.6 m s^{-1}. Samples were then annealed at 150°C for 10 min to dry and remove the residual PVP from the AuNPs. Patterning of the AuNPs onto the polyimide is possible, using the flexographic printing technique, through the use of patterned printing plates. A short length of tin-coated copper wire was then attached to the carbon electrode in an area where no AuNPs had been printed to allow an electrical connection between a sample and a potentiostat. The wire was attached using a conductive epoxy. Paraffin wax was then used to define a small window (~2 mm²) in the AuNP-coated carbon region for functionalisation. Cysteamine and glutaraldehyde were treated onto the AuNPs prior to the functionalisation of GOx. The sensing window was treated with 5 µl of 20 mM cysteamine for 30 min after which it was washed with DI water and dried using N_2 gas. The sensing window was then exposed to 5 µl of 4% glutaraldehyde solution for 30 min and once again washed and dried as previous. The final step in the electrode preparation was to introduce GOx to the sensing window, where 5 µl of 7 mg ml⁻¹ GOx was added to the window and left overnight to ensure its attachment to the electrode surface. The window was washed thoroughly with PBS and dried under N_2. The printed carbon-AuNP-GOx electrode was then ready for electrochemical experiments. A schematic diagram of the working electrode fabrication process is shown in Figure 2.

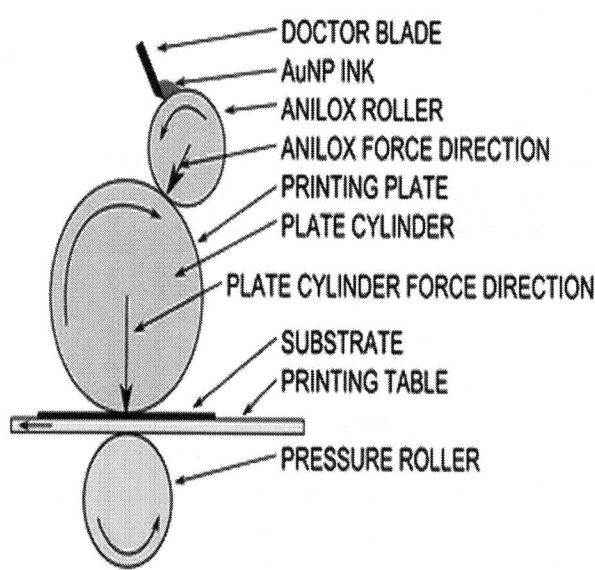

Figure 1: Schematic diagram of a flexographic printer. Ink is transferred to the printing plate from the anilox roller at a rate controlled by the anilox surface features and the doctor blade. Ink is then printed onto the substrate from the printing plate in a continuous roll-to-roll manner.

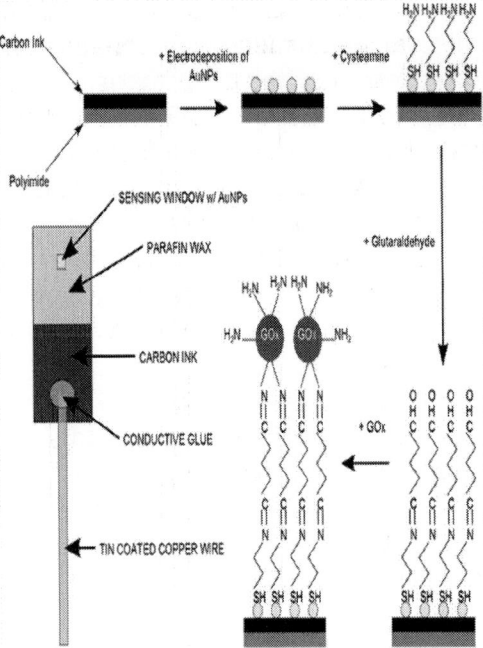

Figure 2: Schematic diagram showing the functionalisation scheme. The scheme shows the major steps in the functionalisation scheme for AuNPs on carbon electrode. Starting with the printed carbon electrode AuNPs are printed and functionalized with cysteamine, glutaraldehyde and glucose oxidase, also included is a schematic of a fully prepared electrode for biosensing (left).

Characterisation

The morphology of the AuNPs coated onto the carbon substrate was studied using a Hitachi S-4800 scanning electron microscope (SEM) (Hitachi High-Technologies Corporation, Tokyo, Japan) at 20 kV. The SEM was equipped with an Oxford Instruments energy dispersive X-ray (EDX) detector (Oxford Instruments, Oxfordshire, UK) which was subsequently used to determine the chemical composition of the printed AuNP samples. Scanning auger microscopy (SAM) and auger electron spectroscopy (AES) were carried out using an Omicron NanoSAM detector (Omicron NanoTechnology GmbH, Taunusstein, Germany) with an accelerating voltage of 5 kV, 1 nA beam current and a 90-μm aperture. Electrochemical measurements were carried out using a CompactStat (Ivium Technologies, Eindhoven, The Netherlands). All electrochemical experiments were conducted using a three-electrode system which comprised of a printed carbon-AuNP-GOx working electrode, a gold-wire counter electrode and a Ag/AgCl reference electrode. The three-electrode system was always orientated in the same manner to avoid any variation from sample to sample. All electrochemical experiments were performed at room temperature inside a Faraday cage, PBS was used as the supporting electrolyte, and experiments were conducted under constant stirring. The magnetic stirring provided the electrolyte solution with sufficient conductive transport. Chronoamperometry was performed at +0.8 V with an interval time of 1 s in an air-saturated environment and used for glucose detection. The current was allowed to settle to a constant value prior to any glucose additions.

RESULTS AND DISCUSSION

AuNP ink formulations were tested on printed carbon electrodes fabricated as described previously. Many different ink formulations and printing parameters were explored to find the best approach. The evaporation rate and polymer content of the ink is of paramount importance. In early attempts, high IPA percentages of 80% to 95% were adopted. It was seen that the solvent would evaporate too quickly resulting in a large loss of gold on the printing plate. Addition of extra polymer to the ink was also tested aiming to improve dispersion stability. This resulted in an increased viscosity, but also decreased wetting of the carbon surface with the AuNP ink. This resulted in visibly non-uniform deposition of polymer at the surface. With 70% IPA and 30% water, the printing of the AuNP ink was relatively uniform and the AuNP distribution on the carbon surface was suitable for sensor testing. The printing force also had an effect on the quality of the printing. It was observed that high printing forces would cause the ink to spread too much, and low printing forces would result in blobs of ink at the electrode surface. The printing force was optimised to 125 N as this gave the most consistent and evenly distributed matrix of AuNPs on the surface of the carbon electrode. It was found that rapid addition of ice-cold $NaBH_4$ produced smaller nanoparticles than synthesis done with room temperature $NaBH_4$ added more slowly. With the synthesis parameters described above, an ink with the desired characteristics, such as small AuNPs and good viscosity, to hold the nanoparticles in suspension was achieved.

Figure 3a shows the contact angle of a water droplet, which is shown as a control, after drop casting onto a printed carbon electrode. The droplet shows a contact angle of 135° indicating poor wetting of the electrode surface. In order to investigate the wetting issues seen by ink formulations with high PVP concentration, an additional 0.5 g of PVP was added to our ink and drop cast onto the printed carbon surface. Figure 3b shows the contact angle at this concentration to be 42° and indicates better wetting than water. However, the surface of the printed carbon was not completely wetted by the ink with this formulation. This resulted in inconsistent coverage where, after drying, some areas contained high densities of AuNPs and others contained low densities. This can occur throughout the printing process and can have a detrimental effect on the quality of the printed electrodes. Figure 3c shows a contact angle image of the optimised AuNP ink developed in this work which was used in the fabrication of the working electrode. The image shows the ink wets the surface well and has a very small contact angle of 6°. This demonstrates that the ink has the required wetting properties to provide consistency and uniformity in the deposition of AuNPs on the substrate.

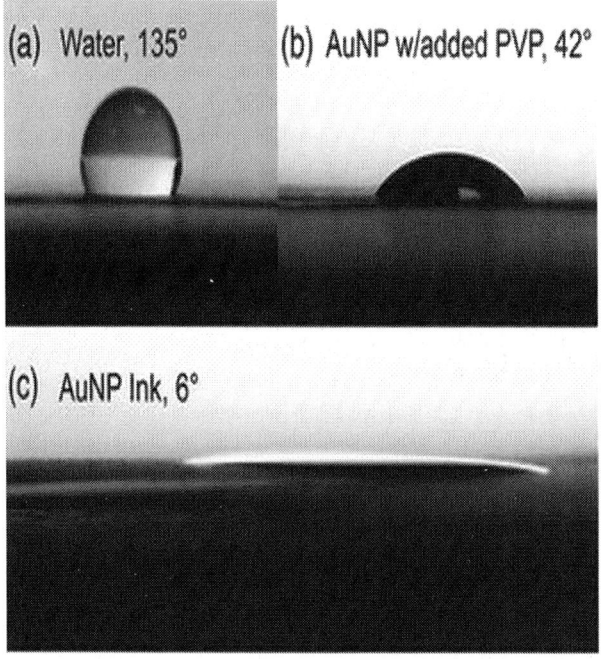

Figure 3: Images showing the contact angles of inks and for comparison, water. Images showing contact angles for (a) water, (b)AuNP ink + 0.5 g PVP and (c) AuNP ink, after drop casting onto printed carbon substrates. The images show contact angles of (a) 135°, **(b)** 42°, and (c) 6°, respectively, showing improved wetting by the AuNP ink.

Figure 4 shows SEM images of printed AuNPs on carbon-polyimide substrate. Figure 4a demonstrates the selective patterning of AuNPs using the flexographic printing technique, producing line widths of <120 μm from a printing plate with a 100-μm width; the lines also displayed low edge distortion, and line width shows a linear relationship with printing force. The printed honeycomb pattern has well-defined edges, and this intricate pattern highlights the capabilities of the technique to perform selective patterning of the AuNPs during device fabrication. This is a highly desirable characteristic as it allows for areas of a substrate to be separated from each other. This is particularly applicable to full sensor fabrication as it allows the AuNP electrode to be electrically isolated from other electrodes on the same substrate. Figure 4b shows a uniform distribution of AuNPs on the carbon surface. There are some areas of agglomeration, but this is not deemed significant enough to have a detrimental effect on the biosensing capabilities of the electrode. Figure 4c is a 50-kx magnification image of AuNPs printed onto the substrate. The image shows limited agglomeration of the AuNPs, and the distribution of the particles is relatively uniform, which is typical of the entire printed area. Figure 4d is a 100-kx magnification image of the AuNPs, it can be seen that the size of the majority of nanoparticles is less than 60 nm.

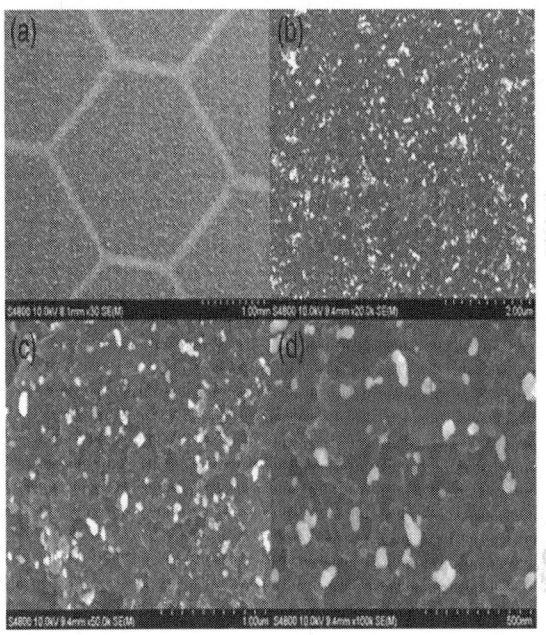

Figure 4: SEM image of printed AuNP ink on carbon electrodes showing the pattering of the particles. (a) SEM characterisation shows well-defined patterning of the AuNP ink in a printed honeycomb pattern on carbon electrodes, at low magnification (30×). Higher magnification images(b) 20, (c) 50 and (d) 100 kx show the even distribution of the AuNPs on the surface.

Particle size analysis was carried out using MATLAB particle size distribution software on a 50-kx magnification SEM image of printed AuNPs. Figure 5 is a histogram showing size distribution of printed AuNPs on a carbon-polyimide substrate after flexographic printing using the parameters previously described. The vast majority of particles are under 60 nm, which is ideal for high-sensitivity biosensing applications due to the very large surface to volume ratio of these nanoparticles, and they favour the immobilisation of enzymes [20]. There are a very small number of particles over 100 nm, and these large particles are likely to be agglomerations of smaller particles. Due to the very small number of large particles, we can assume that they will not have a significant effect on the performance of the sensor.

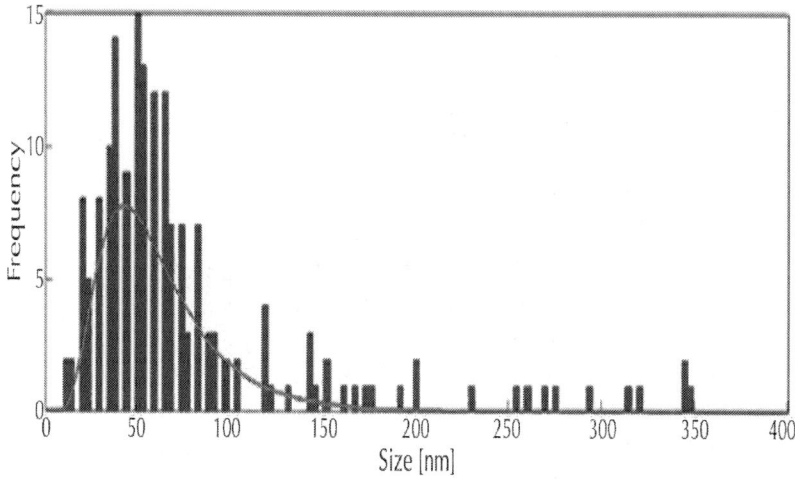

Figure 5: Histogram showing AuNP size distribution on carbon electrode. After flexographic printing of the AuNP ink, particle size analysis was carried out on a 50-kx magnification SEM image of the particles. This analysis shows that the majority of the particles are less than 60 nm in size.

A simple thermogravimetric process was carried out to assess the presence of PVP after post-annealing of the printed surface. AuNP ink was drop cast onto a substrate and dried at room temperature overnight. The sample was then weighed, annealed at 150°C for 10 min and re-weighed yielding a 62% drop in weight which corresponds to the loss of polymer from the surface. This correlates with results seen by Cui et al. where they saw a total mass loss of 57% attributed to the loss of PVP [21]. This post-annealing step to remove PVP is important for the subsequent functionalization of the electrode for biosensing applications.

A combination of SAM and AES analyses was used to investigate the printed AuNPs at the carbon electrode. Figure 6a shows a SAM image of the printed AuNPs on carbon electrode. The scan was performed by fixing the detector to the energy of the Au3 MNN transition and raster scanning across the scan window. The high-intensity regions correspond to gold on the substrate surface, with the darker areas representative of the printed carbon (no AuNPs). Point AES measurements were performed on and off the AuNPs as shown in Figure 6b. The red line corresponds to point spectra performed on the AuNP whereas the black line shows point spectra on bare carbon. AES performed at the bare carbon showed only one observable peak in the survey spectra at 262.5 eV, corresponding to the C1 KLL auger transition. Spectra on the AuNPs, however, show a distinctive peak at 2,016 eV, characteristic of the Au3 MNN transition. The presented AES spectra in Figure 6b have been differentiated using a 2-eV Savitzky-Golay smoothing width to enhance the signal to noise ratio without losing peak information. The AES measurements suggest that the samples comprise of only carbon and AuNPs on the surface as there are no observable peaks other than those from carbon and gold transitions. This is further supported by the energy dispersive X-ray (EDX) spectrum as shown in Figure 7. There are two large peaks which are attributed to carbon and gold at the printed AuNP-carbon electrode. The large gold peak indicates the presence of substantial gold on the carbon surface. The inset shown in Figure 7 highlights the nitrogen and oxygen peaks before and after annealing, there is a significant drop in the number of counts and this is due to the loss of PVP from the electrode surface which contains nitrogen and oxygen groups in its structure. It is important to remove PVP as it could have a detrimental effect on the performance of the device. The result shows that the printing techniques developed and employed in this work are capable of producing surfaces with AuNPs that are uniformly distributed onto the substrate, which is ideal for the fabrication of devices such as biosensors.

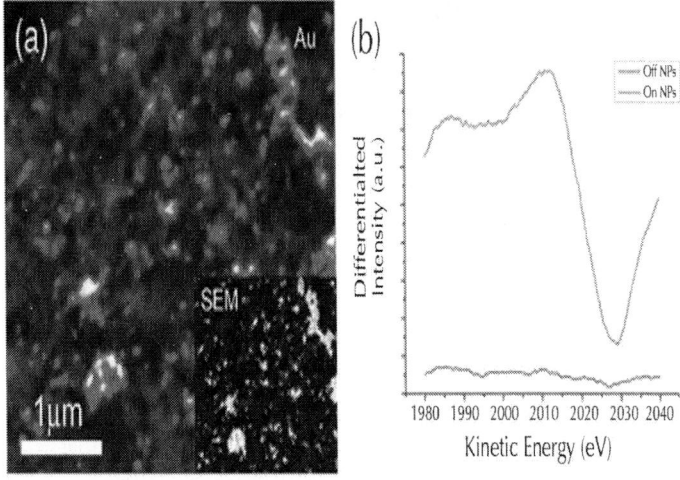

Figure 6: Scanning auger microscopy analysis of printed AuNPs on carbon electrode. (a) Scanning auger microscopy image of AuNPs using peak intensity from the centre of the Au peak (inset shows SEM image of corresponding area). (b) Differentiated spectra taken from areas on the AuNPs (red) and off of the AuNPs (black).

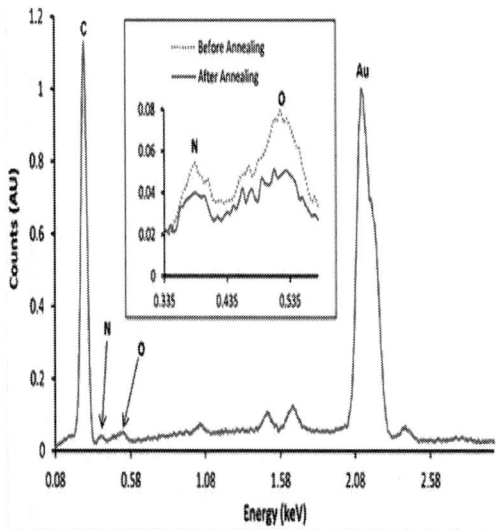

Figure 7: EDX spectrum of printed AuNPs at carbon electrode before and after annealing. Energy dispersive X-ray analysis of the AuNPs printed onto a

carbon electrode shows carbon, gold, nitrogen and oxygen. The inset shows nitrogen and oxygen peaks before (red) and after (blue) annealing, indicating a reduction in nitrogen and oxygen peaks due to the annealing process.

AuNPs provide a biocompatible environment for the use of enzymes, such as GOx. In studies using such nanoparticles, it has been shown that the activity of GOx can be enhanced [22]. Figure 8 shows a chronoamperometric graph for a typical printed polyimide-carbon-AuNPs-GOx electrode fabricated as described previously. It can be seen from the graph that there is a clear, rapid response to glucose additions. The sensor displays a 2.1-nA step for a 0.01-mM addition. The sensor exhibited 18.4-nA steps for ten subsequent 0.1-mM glucose additions and a large 72.2-nA step for a 0.5-mM glucose addition. The clear steps and fast response exhibited are highly desirable in PoC diagnostic devices.

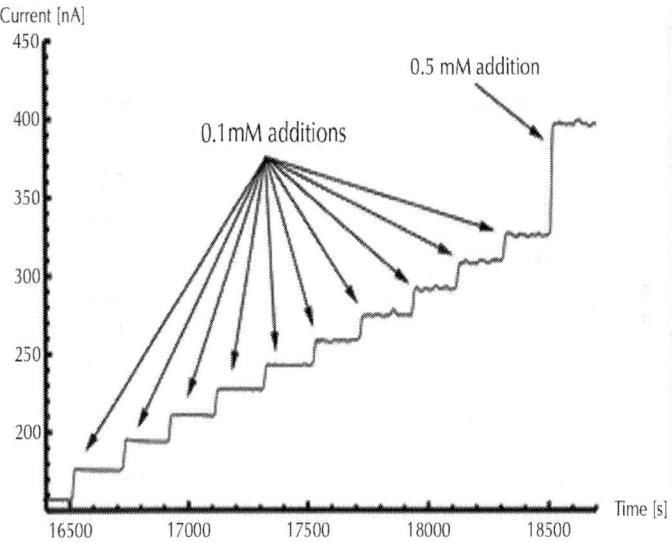

Figure 8: Chronoamperometric graph showing fabricated polyimide-carbon-AuNP-GOx electrode response to glucose additions. After functionalisation, samples were tested for the electrochemical detection of glucose. These samples where tested with glucose concentration step increments of 0.1 and 0.5 mM as detailed in the figure.

Figure 9 is the calibration curve for the glucose sensor. The graph shows a very good linear response to glucose additions, resulting in an

R^2 value of 0.9979. The linear range of the sensor was from 0.01 to 1.5 mM. After this point, the device began to saturate which was indicated by a plateau beginning to form with additions after 1.5 mM. The sensor displayed a high sensitivity of 5.52 μA mM^{-1} cm^{-2} with a detection limit of 26 μM. The limit of detection was calculated using the following formula:

$$LoD = 3RSD_B/I_{mM}$$

(RSD_B is the relative standard deviation of blank signal, and I_{mM} is the current per millimole).

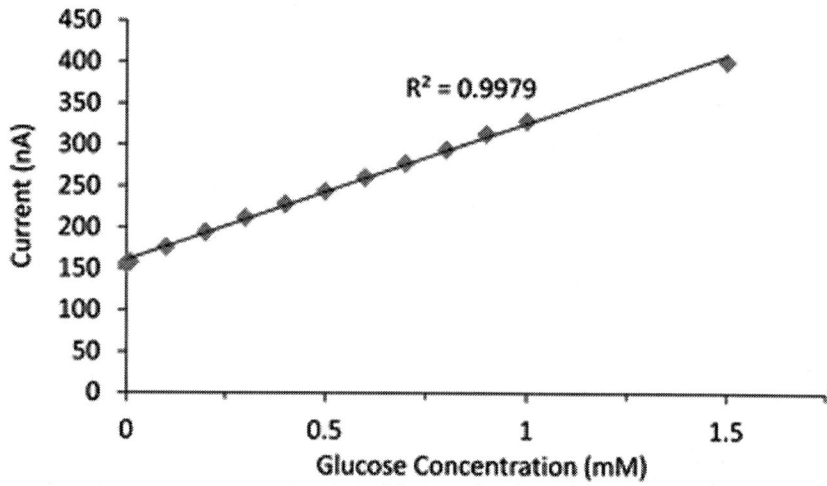

Figure 9: Calibration curve for the polyimide-carbon-AuNP-GOx electrode responding to glucose. Plot of steady state current against concentration of glucose taken from the chronoamperometric data for a polyimide-carbon-AuNP-GOx electrode. The graph shows the linear response of the electrode for glucose additions up to 1.5 mM.

The sensitivity of this device is similar to those previously reported for AuNP glucose sensors functionalised through similar methods [10], [11]. However, the printed device shown here has demonstrated the use of a fabrication technique which holds the potential for mass production.

This work has demonstrated the use of flexographic printing for the patterning of AuNPs onto electrode surfaces for biosensing applications. The main issue with previous work carried out on AuNPs is that much of the fabrication technique is not transferable to mass production. Glucose sensing is just one of many potential applications of this technology. The use of flexographic printing can significantly reduce the production cost of the devices as it allows the fabrication of vast numbers of devices at relatively low cost. This is due to the fast printing process with great accuracy allowing direct patterning of nanomaterials onto a substrate. This can be compared to methods employed by others whereby the attachment of AuNPs to the surface of the electrode takes a 6-h immersion in an AuNP solution [10], while flexographic printing would take seconds to print potentially hundreds of samples. Flexographic printing offers very high throughput as compared with other printing techniques, such as screen and inkjet printing. The technique is able to perform roll-to-roll printing at a speed of up to 300 m min^{-1}. Inkjet printing is significantly slower and also suffers from problems such as blockage of the fine printing head and 'coffee-ring' effect [23]. It is clear that the printing of AuNPs via flexographic printing allows a high throughput of devices, and the sensitivities achieved on our printed AuNP glucose sensors are comparable to those seen through other fabrication methods.

CONCLUSIONS

This work has demonstrated the direct patterning of AuNPs on substrates using flexographic printing which is ideal for volume production at a relatively low cost. Results showed that the printed AuNPs are uniformly distributed on the carbon electrode using an optimised ink formulation and printing parameters. AuNP-based glucose sensors developed using the technique display a rapid response to the addition of glucose and have a sensitivity of 5.52 µA mM^{-1} cm^{-2} with a low detection limit of 26 µM. High throughput, low-cost production of AuNP-based biosensors is highly viable, and the technology can be transferred to a wide range of biosensing applications.

AUTHORS' CONTRIBUTIONS

JB and KST designed the experiments and wrote the manuscript in close collaboration with other authors. JB, DD and KST analysed the data. JB and CMF performed, characterised and optimised the printing processes. JB and JSL performed biosensing experiments. NAS performed AES and SAM experiments. All authors discussed the results and approved the final version of the manuscript.

ACKNOWLEDGEMENTS

This work was financially supported by the Welsh Government (project reference no. HE 09 15 1012).

REFERENCES

1. Ghosh P, Han G, De M, Kim CK, Rotello VM: Gold nanoparticles in delivery applications. *Adv Drug Deliv Rev* 2008, 60:1307-15. doi:10.1016/j.addr.2008.03.016

2. El-Sayed IH, Huang XH, El-Sayed MA: Surface plasmon resonance scattering and absorption of anti-EGFR antibody conjugated gold nanoparticles in cancer diagnostics: applications in oral cancer. *Nano Lett* 2005, 5:829-34.doi:10.1021/nl050074e

3. Murphy CJ, Gole AM, Stone JW, Sisco PN, Alkilany AM, Goldsmith EC: Gold nanoparticles in biology: beyond toxicity to cellular imaging.*Acc Chem Res* 2008, 41:1721-30.doi:10.1021/ar800035u

4. Cao YC, Jin RC, Nam JM, Thaxton CS, Mirkin CA: Raman dye-labeled nanoparticle probes for proteins.*J Am Chem Soc* 2003, 125:14676-7.doi:10.1021/ja0366235

5. Wang J: Electrochemical biosensing based on noble metal nanoparticles.*Microchim Acta* 2012, 177:245-70.doi:10.1007/s00604-011-0758-1

6. Eustis S, El-Sayed MA: Why gold nanoparticles are more precious than pretty gold: noble metal surface plasmon resonance and its enhancement of the radiative and nonradiative properties of nanocrystals of different shapes.*Chem Soc Rev* 2006, 35:209-17. doi:10.1039/B514191E

7. Doria G, Conde J, Veigas B, Giestas L, Almeida C, Assuncao M, et al.: Noble metal nanoparticles for biosensing applications *Sensors (Basel)* 2012, 12:1657-87.doi:10.3390/s120201657

8.   Feng D, Wang F, Chen Z: Electrochemical glucose sensor based on one-step construction of gold nanoparticle-chitosan composite film.*Sens Actuators B Chem* 2009, 138:539-44 doi:10.1016/j.snb.2009.02.048

9.   Jena BK, Raj CR: Enzyme-free amperometric sensing of glucose by using gold nanoparticles.*Chemistry* 2006, 12:2702-8. doi:10.1002/chem.200501051

10. Zhang SX, Wang N, Yu HJ, Niu YM, Sun CQ: Covalent attachment of glucose oxidase to an Au electrode modified with gold nanoparticles for use as glucose biosensor.*Bioelectrochemistry* 2005, 67:15-22.doi:10.1016/j.bioelechem.2004.12.002

11. Zhang SX, Wang N, Niu YM, Sun CQ: Immobilization of glucose oxidase on gold nanoparticles modified Au electrode for the construction of biosensor.*Sens Actuators B Chem* 2005, 109:367-74.doi:10.1016/j.snb.2005.01.003

12. Mr. Rohan. Point-of-care diagnostic market worth $27.5 billion by 2018. http://www.prnewswire.com/news-releases/point-of-care-diagnostic-marketworth-275-billion-by-2018-274885521.html (2014). Accessed 7 Jan 2015.

13. Nagata R, Yokoyama K, Clark SA, Karube I: A glucose sensor fabricated by the screen printing technique.*Biosens Bioelectron* 1995, 10:261-7.doi:10.1016/0956-5663(95)96845-P

14. Setti L, Fraleoni-Morgera A, Ballarin B, Filippini A, Frascaro D, Piana C: An amperometric glucose biosensor prototype fabricated by thermal inkjet printing.*Biosens Bioelectron* 2005, 20:2019-26.doi:10.1016/j.bios.2004.09.022

15. Soltman D, Subramanian V: Inkjet-printed line morphologies and temperature control of the coffee ring effect.*Langmuir* 2008, 24:2224-31.doi:10.1021/la702684

16. Ahmad M, Pan C, Luo Z, Zhu J: A single ZnO nanofiber-based highly sensitive amperometric glucose biosensor.*J Phys Chem C* 2010, 114:9308-13.doi:10.1021/jp102505g

17. You X, Park J, Jang Y, Kim S, Pak JJ, Min NK. Enzymatic glucose biosensor based on porous ZnO/Au electrodes. IEEE Int Conf Nano/Mol Med Eng. 2010:56–9. doi:10.1109/NANOMED.2010.5749805.

18. Shulga O, Kirchhoff JR: An acetylcholinesterase enzyme electrode stabilized by an electrodeposited gold nanoparticle layer.*Electrochem Commun* 2007, 9:935-40.doi:10.1016/j.elecom.2006.11.021

19. Bharathi S, Nogami M: A glucose biosensor based on electrodeposited biocomposites of gold nanoparticles and glucose oxidase enzyme.*Analyst* 2001, 126:1919-22.doi:10.1039/B105318N

20. Zhou K, Zhu Y, Yang X, Luo J, Li C, Luan S: A novel hydrogen peroxide biosensor based on Au-graphene-HRP-chitosan biocomposites.*Electrochim Acta* 2010, 55:3055-60. doi:10.1016/j.electacta.2010.01.035

21. Cui W, Lu W, Zhang Y, Lin G, Wei T, Jiang L: Gold nanoparticle ink suitable for electric-conductive pattern fabrication using in ink-jet printing technology.*Colloids Surf A Physicochem Eng Asp* 2010, 358:35-41.doi:10.1016/j.colsurfa.2010.01.023

22. Kaushika A, Khan R, Solanki PR, Pandey P, Alem J, Ahmad S, *et al.*: Iron oxide nanoparticles-chitosan composite based glucose biosensor.*Biosens Bioelectron* 2008, 24:676-83.doi:10.1016/j.bios.2008.06.032

23. Kitsomboonloha R, Baruah S, Myint MTZ, Subramanian V, Dutta J: Selective growth of zinc oxide nanorods on inkjet printed seed patterns.*J Cryst Growth* 2009, 311:2352-8.doi:10.1016/j.jcrysgro.2009.02.02

Chapter 7

Biosensor for Human IgE Detection Using Shear-mode FBAR Devices

Ying-Chung Chen[1], Wei-Che Shih[1], Wei-Tsai Chang[1],
Chun-Hung Yang[1], Kuo-Sheng Kao[2],
and Chien-Chuan Cheng[3]

[1]Department of Electrical Engineering, National Sun Yat-Sen University, Kaohsiung 80424, Taiwan

[2]Department of Computer and Communication, Shu-Te University, Kaohsiung 82445, Taiwan

[3]Department of Electronic Engineering, De Lin Institute of Technology, Taipei 23654, Taiwan

ABSTRACT

Film bulk acoustic resonators (FBARs) have been evaluated for use as biosensors because of their high sensitivity and small size. This study fabricated a novel human IgE biosensor using shear-mode FBAR devices with c-axis 23°-tilted AlN thin films. Off-axis radio frequency (RF) magnetron sputtering method was used for deposition of c-axis 23°-tilted AlN thin films. The deposition parameters were adopted as working pressure of 5 mTorr, substrate temperature of 300°C, sputtering power of 250 W, and 50 mm distance between off-axis and on-axis. The characteristics of the AlN thin films were investigated by X-ray diffraction and scanning electron microscopy. The frequency response was measured with an HP8720 network analyzer with a CASCADE probe station. The X-ray diffraction revealed (002) preferred wurtzite structure, and the cross-sectional image showed columnar structure with 23°-tilted AlN thin films. In the biosensor, an Au/Cr layer in the FBAR backside cavity was used as the detection layer and the Au surface was modified using self-assembly monolayers (SAMs) method. Then, the antigen and antibody were coated on biosensor through their high specificity property. Finally, the shear-mode FBAR device with k_t^2 of 3.18% was obtained, and the average sensitivity for human IgE detection of about 1.425×10^5 cm^2/g was achieved.

BACKGROUND

In recent years, piezoelectric materials have been used in surface acoustic wave (SAW) resonators[1]-[5] and film bulk acoustic wave resonators (FBARs) [6]-[10] because of their low cost, low weight, and good reproducibility. However, the SAW resonator has high insertion loss and poor power handling capability. Hence, this study evaluated the potential applications of FBARs for biosensors because of their advantages, including low insertion loss, good power handling, and small size. The FBAR devices were constructed by a piezoelectric layer sandwiched between two electrodes and attached to substrate with backside cavity. Piezoelectric materials such as zinc oxide (ZnO) and aluminum nitride (AlN) have been used in FBAR devices for various applications [11]-[13] owing to their high acoustic velocity, better quality factor, and high electromechanical coupling coefficient. Besides, the piezoelectric materials of ZnO and AlN can be combined with silicon technologies in semiconductor fabrication processes [14],[15]. Moreover, the acoustic velocity of AlN is 10,400 m/s, and it suits application for FBAR devices.

The acoustic wave of a FBAR has two transmittance modes: longitudinal mode and shear mode. In shear mode, acoustic wave energy does not dissipate in a liquid environment [16]. The backside cavity of FBAR can be used as the detection area for adsorbent matter. Under a mass loading, a frequency shift would be resulted in the frequency response of a FBAR [17]. The analysis methods were used for biosensor in liquid and tiny mass detection in air through the shear mode and longitudinal mode, respectively. Thus, FBAR devices were fabricated and constructed to evaluate their potential use in biosensors.

According to the medicine journal report, it is estimated that as many as 1.4 billion people of allergy[18]. Hence, the marketable merit of anti-allergic agent is calculated to be 20 billion USD dollars[19]. The conventional detecting allergy methods focus on testing the concentration of immunoglobulins E (IgE) in human serum. The IgE in human immune system was used to resist exterior germs and virus, but overreactions of the human immune system can cause allergies. Furthermore, the traditional detecting allergy methods have some disadvantages such as time-consuming detecting process and large size and expensive detecting instrument [20]. Therefore, this investigation focuses on micro allergic sensor devices, due to the advantages, such as small size, low-cost, fast detecting process, etc. Besides, the apparatus for evaluating the FBAR-based sensor devices is shown in Figure 1a, b.

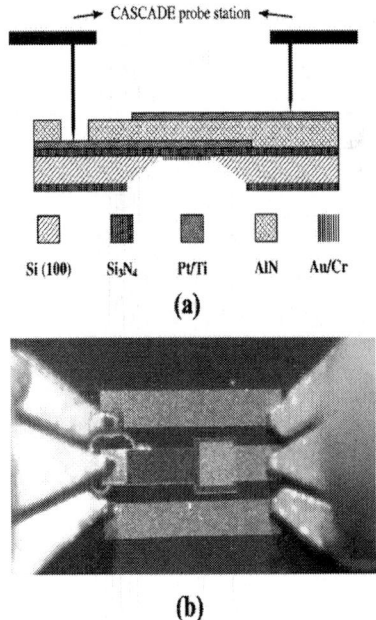

Figure 1: The apparatus for evaluating the FBAR-based sensor devices. (a) Schematic cross section view. (b) Front view.

METHODS

Fabrication of FBAR Devices

In this study, the FBAR devices for biosensors application were fabricated. Figure 2 showed the processes used to fabricate the shear-mode FBAR devices. The silicon nitride (Si_3N_4) was deposited on both sides of Si wafer by low-pressure chemical vapor deposition (LPCVD) as the supporting layer for the FBAR devices. The bottom electrodes, piezoelectric thin films, and top electrodes sandwiched structure is patterned by the photolithography process using four masking processes. The titanium (Ti) and platinum (Pt) layers were deposited on Si_3N_4/Si structure as bottom electrodes by a dual-gun DC sputtering system using 99.995% pure targets combined with first mask and lift-off method. The distance between target and substrate was fixed at 50 mm. As the base pressure was pumped down to 1×10^{-6} Torr, the film growth was carried out with working pressure of 3 and 1 mTorr for Ti and Pt, respectively. Then the high-quality AlN piezoelectric thin films were deposited on Pt/Ti layer using reactive radio frequency (RF) magnetron sputtering with off-axis deposition method. The Al target was 99.9995% pure, and the distance between target and substrate was fixed at 50 mm. As the base pressure was pumped down to 5×10^{-7} Torr, the sputtering conditions were set as working pressure of 5 mTorr, substrate temperature of 300°C, sputtering power of 250 W, and an off-axis to on-axis distance of 50 mm. To expose the bottom electrodes for electrical contact, AlN was wet-etched with 2.38% tetramethylammonium hydroxide (TMAH) using a second mask at room temperature. The top electrode can be obtained by the third patterning process after Pt/Ti was deposited on the AlN thin films. Finally, the backside of the structure was etched by combining the fourth mask and a 30% KOH solution to form the detection area. Therefore, the fabrication of the FBAR devices was then completed.

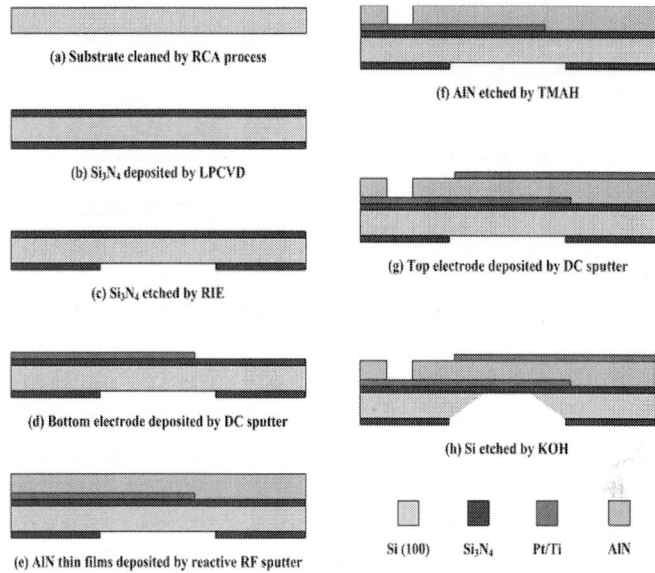

(a) Substrate cleaned by RCA process

(b) Si₃N₄ deposited by LPCVD

(c) Si₃N₄ etched by RIE

(d) Bottom electrode deposited by DC sputter

(e) AlN thin films deposited by reactive RF sputter

(f) AlN etched by TMAH

(g) Top electrode deposited by DC sputter

(h) Si etched by KOH

Si (100)   Si₃N₄   Pt/Ti   AlN

Figure 2: The fabrication steps of FBAR devices.

Characteristics Measurement

The characteristics of AlN thin films, including crystalline properties, preferred orientation, and cross-sectional morphologies were examined. The crystalline properties and preferred orientation of the AlN thin films were determined by X-ray diffraction scanning between 20° and 60° using a Siemens D5000 (Munich, Germany) with CuKα radiation. The surface morphologies and cross sections of AlN thin films were observed by field-emission scanning electron microscope (FESEM, JEOL-6700; JEOL Ltd., Akishima-shi, Japan). Finally, the frequency responses of FBAR devices with the biosensors were measured by HP8720 network analyzer.

FBAR Devices for Biosensor Applications

For biosensor applications of the FBAR, Au/Cr thin films were deposited in the backside cavity of FBAR devices as the detection layer using a dual-gun DC sputtering system as shown in Figure 3. In the self-assembly monolayers (SAMs) method, the Au surface was modified by adsorption of thiolate $(CH_3 (CH_2)_n SH)$.

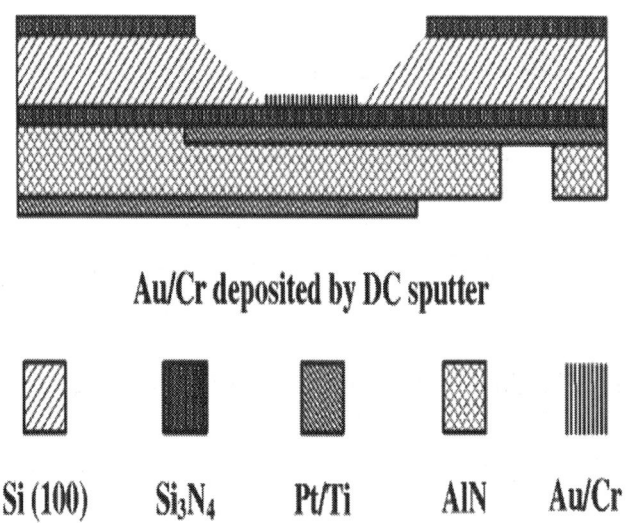

Au/Cr deposited by DC sputter

Si (100)   Si₃N₄   Pt/Ti   AlN   Au/Cr

Figure 3: The schematic diagram of a biosensor.

The SAMs method was performed as follows:

Step (1): Use oxygen plasma process for Au surface cleaning.

Step (2): Inject cysteine solution (R.T., 1 h).

Step (3): Inject deionized (DI) water and dry using N_2 gas.

Step (4): Inject glutaraldehyde solution (2.5%, R. T., 1 h).

Step (5): Inject DI water and dry with N_2 gas.

Then, the surface modification of FBAR devices were accomplished by the SMAs method. In the biosensors, human IgE was detected by using a coating process to detect antibody with antigen because of the high specificity between antigen and antibody. Hence, the coating process was performed as described in the literatures as follows [21]-[24]:

Step (1): Wash with 200 µl phosphate-buffered saline (PBS) solution three times.

Step (2): Dip 200 µl diluted mouse anti-human IgE antibody (37°C, 2 h).

Step (3): Inject 200 µl, Tween-20 wash buffer three times.

Step (4): Inject 200 µl, 10 wt.% bovine serum albumin (BSA) solution (37°C, 0.5 h).

Step (5): Inject 200 µl, Tween-20 wash buffer three times.

Step (6): Inject 200 µl, diluted human IgE antigen with 0.707 µg/ml concentration.

In the backend process of step (2) to step (6), the sample were cleaned using DI water to remove excess liquid and then dried with N_2 gas. Figure 4 schematically depicts the IgE antigen/IgE antibody/glutaraldehyde/the integrated cystamine SAMs multilayer [20].

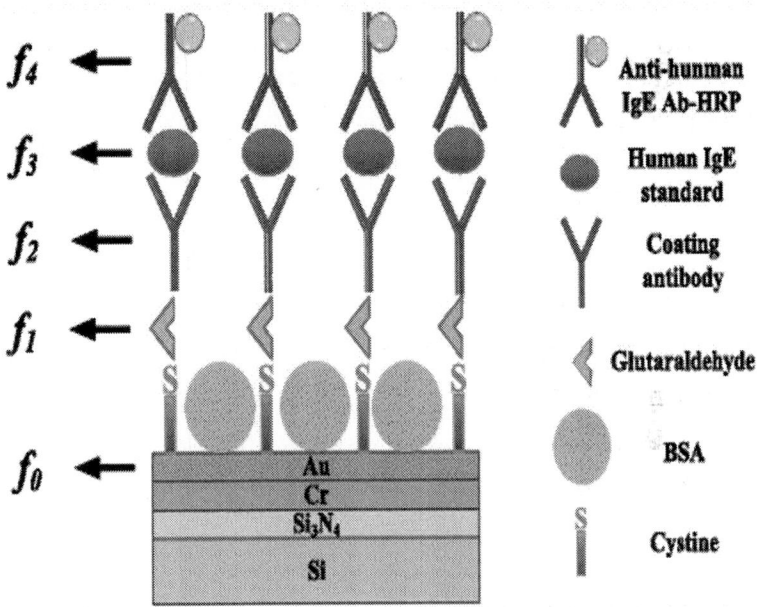

Figure 4: The schematic diagram for the integration of cystamine SAM, glutaraldehyde, IgE antibody and antigen multilayer.

After the above SAMs and coating processes, the frequency response was measured before and after anti-human IgE antibody linked with the human IgE antigen. Finally, the sensitivity (S_m) of the biosensor was calculated using the following equation:

$$S_m = \lim_{\delta m \to 0} \left(\frac{\delta f}{f}\right) \left(\frac{1}{\delta m}\right),$$

where δm is the loading mass (9.1875 ng/cm^2) and δf is the variation of the resonate frequency. Finally, the sensitivities of FBAR devices for human IgE detection were investigated.

RESULTS AND DISCUSSION

Structural and Morphological Properties of AlN Thin Films

A highly c-axis orientation is the ideal piezoelectric property of a FBAR device. According to the literature, a 34.5° c-axis tilted piezoelectric thin film in FBAR device exits strongly shear-mode transmittance [25]. The optimized sputtering conditions for 23° c-axis tilted highly textured AlN thin films were obtained in our previous report [26], those were working pressure of 5 mTorr, substrate temperature of 300°C, sputtering power of 250 W, and the off-axis of 50 mm. Figure 5 shows the c-axis preferred orientation of AlN thin films with small full width at half maximum (FWHM). Besides, Figure 6 shows the cross-sectional images, which reveal columnar with 23°-tilted AlN thin films.

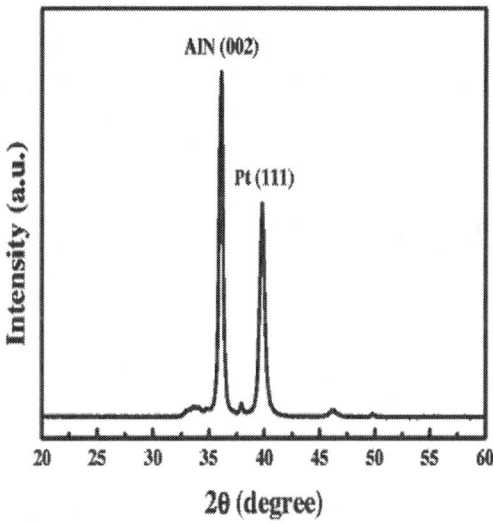

Figure 5: The θ-2θ X-ray scans of the AlN thin film.

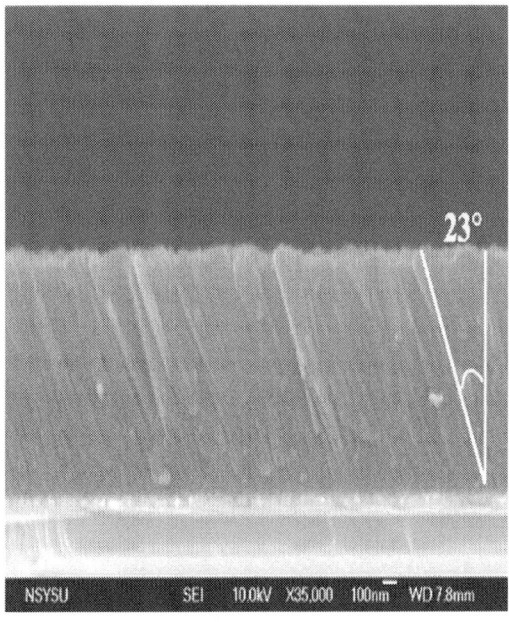

Figure 6: The cross-sectional image of the AlN thin film.

Frequency Responses of Shear-mode FBAR Devices

Figure 7 shows the frequency responses of the FBAR devices with 23°-tilted AlN thin films, in which the longitudinal mode and shear-mode exist at 2.07 (f_L) and 1.175 GHz (f_S), respectively. The ratio of f_L to f_S can be determined from the following relationship:

$$\frac{f_L}{f_S} = \frac{V_L}{V_S} = \frac{\sqrt{\frac{C_{33}}{\rho}}}{\sqrt{\frac{C_{44}}{\rho}}} = \sqrt{\frac{C_{33}}{C_{44}}} = \sqrt{\frac{395 \text{ Gpa}}{118 \text{ Gpa}}} = 1.83,$$

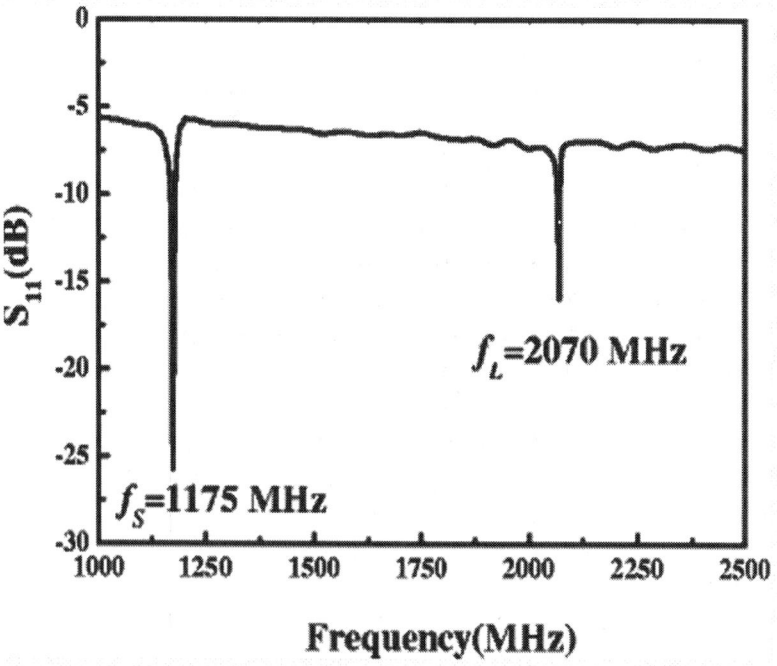

Figure 7: The frequency response of a FBAR device without Au/Cr coatings.

where V_L and V_S are the acoustic velocity, C_{33} and C_{44} are an elastic constant, and ρ is density of the wurtzite AlN. In this study, the practical acoustic velocity of longitudinal mode is 1.76 times than that of the shear mode, which is still consistent with the literature [25],[27]. The electromechanical coupling coefficient (k_t^2) of shear mode is a numerical measurement of the conversion efficiency between electrical and acoustic energy in piezoelectric materials. The k_t2 of the shear mode of the FBAR was calculated to be about 3.18%.

Frequency Responses of Biosensors for Human IgE Detection

The Au/Cr thin films were adopted as detection layer using a dual-gun DC sputtering system, the oxygen plasma process was used to clean the surface of the Au layer in order to improve the hydrophilic properties of the contact area between the bio-drop and Au layer [28]-[32].

Besides, the analysis methods were used for biosensor in liquid and tiny mass detection in air through the shear mode and longitudinal mode, respectively. Figure 8 shows the frequency response of FBAR device in air and liquid environment. The longitudinal mode almost disappeared in liquid environment because of the decrease of quality factor (Q) which reduces the mass resolution substantially, whereas the shear mode maintains high readability. However, the literatures mentioned that the large reflection coefficient of longitudinal mode in solid and liquid interface which is the key factor result in the acoustic wave vanished. Therefore, the shear mode propagating in solid medium maintains its movement through a liquid environment [33]-[35]. The experimental and analytical results indicate that the longitudinal mode is the key indicator to identify the sensing environment, and the shear mode can be exploited in biosensor applications. Hence, FBAR devices with 23°-tilted AlN thin films are suitable for human IgE detection.

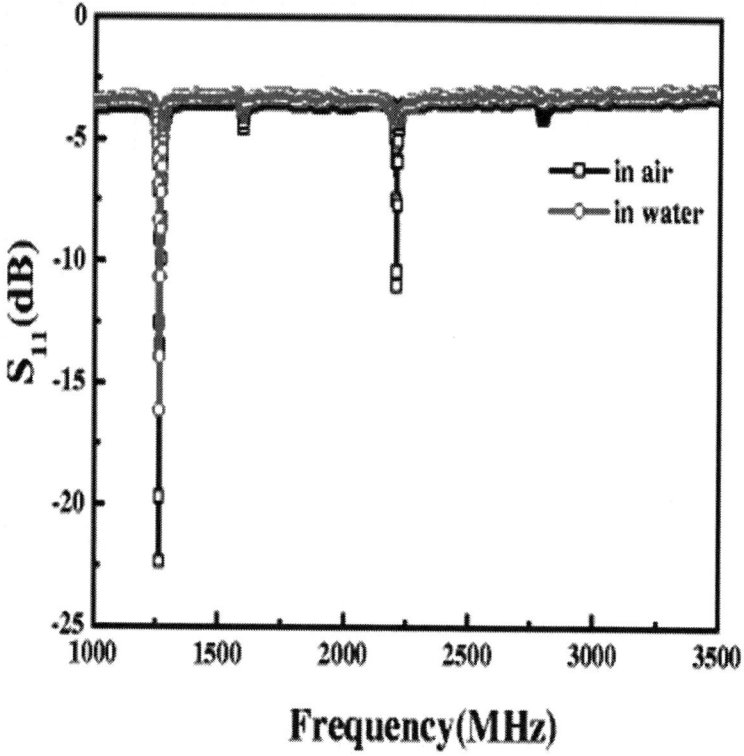

Figure 8: The frequency response of a FBAR device in air and liquid environment.

In this study, two devices of biosensors for human IgE detection were fabricated, and the frequency responses are shown in Figure 9. In Figure 9, f_0, f_1, f_2, f_3, and f_4 are the resonate frequencies of the shear-mode FBAR devices without loading, treated with the SAMs method, combined with the anti-human IgE antibody, linked with the human IgE antigen, and terminated with the anti-human IgE HRP, respectively. The properties of shear-mode FBAR device adopted for the coating mass detection are demonstrated in Figure 9. In some literatures, the resonant frequency were also used to confirm the coating mass adhered on FBAR devices [20],[36].

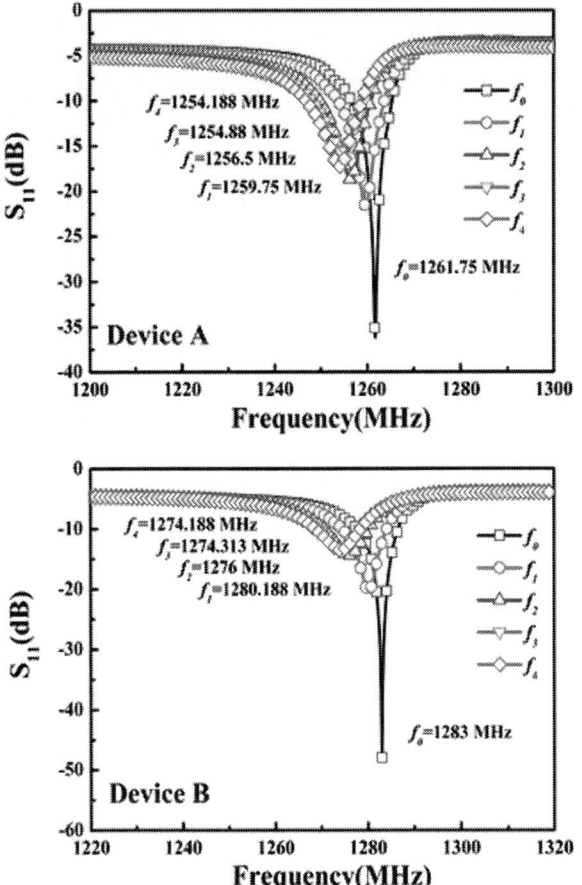

Figure 9: The frequency responses of biosensors for human IgE detection, Device A and Device B

Besides, after repeated testing, the variation of frequency response of the same device exhibited a tiny error of ±0.01%. In the bio-processes, the resonate frequency decreased the range of about 10 MHz which results from the bio-processes effect as SAMs, IgE antibody, IgE antigen, and HRP are added to the biosensor area. Figure 10 shows the variations of resonate frequency step by step from f_0 to f_4. It is confirmed that the matters have mutual bonding when coated on biosensor. However, the standard IgE reagent exist possible error value of ±0.5% in environment according to the official test reports and the enzyme-linked immunosorbent assay (ELISA).

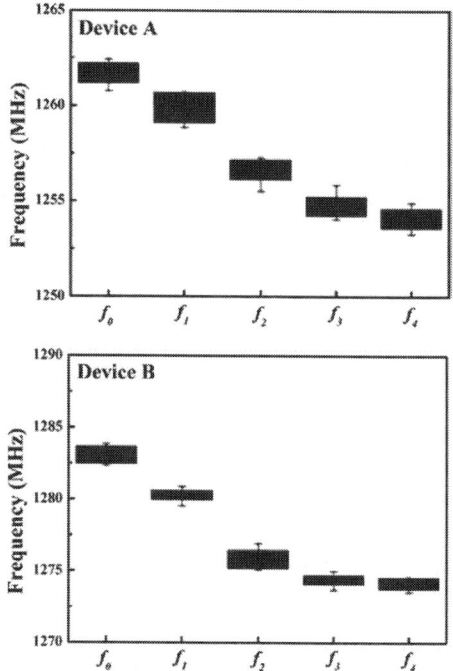

Figure 10: The variations of frequency response step by step from f_0 to f_4, Device A and Device B.

To calculate the sensitivity (S_m) of the shear-mode FBAR devices for human IgE detection, the Sm is calculated $S_m = \lim_{\delta m \to 0}\left(\dfrac{\delta f}{f}\right)\left(\dfrac{1}{\delta m}\right)$ Table 1 shows the calculated sensitivities for human IgE detection of two biosensor devices. The average sensitivity calculated for the shear-mode FBAR devices for human IgE detection was about 1.425×10^5 cm²/g.

Table 1: The frequency shift and sensitivity of biosensors

	Device A	Device B
Frequency shift, δf (MHz)	1.62	1.687
Sensitivity, S_m (cm²/g)	1.41×10^5	1.44×10^5

Chen *et al.*

Chen *et al. Nanoscale Research Letters* 2015 **10**:69, doi:10.1186/s11671-015-0736-3

The results of this study demonstrate that the proposed shear-mode FBAR device is highly promising for use in human IgE detection because of its high sensitivity, small size, low-cost, and rapid reaction process than conventional quartz crystal micro-balance (QCM) [37]-[41].

CONCLUSIONS

This study fabricated shear-mode FBAR devices for biosensor applications. The AlN thin films and Pt/Ti were adopted as the piezoelectric and electrode layers, respectively, in FBAR devices. The AlN thin films were fabricated at a working pressure of 5 mTorr, substrate temperature of 300°C, sputtering power of 250 W, and off-axis of 50 mm. The resulted AlN thin films exhibited a strong c-axis orientation and 23°-tilted. The obtained shear-mode FBAR devices had a frequency response of 1.175 GHz and a k_t^2 of about 3.18%. For biosensor applications, the Au/Cr thin films were deposited on backside cavity of FBAR as bio-detection layer. The SAMs method was used for surface modification of Au thin films. Human IgE was detected by using a coating process to detect antibody with antigen. The average sensitivity for the shear-mode FBAR devices for human IgE detection was about 1.425×10^5 cm^2/g.

AUTHORS' CONTRIBUTIONS

WCS carried out the bulk acoustic resonators studies and drafted the manuscript. YCC, WTC and CHY participated in the design of the study. KSK and CCC conceived of the study and participated in its design and helped to draft the manuscript. All authors read and approved the final manuscript.

ACKNOWLEDGEMENTS

The authors gratefully acknowledge the financial support from the National Science Council, the Republic of China (NSC Grant Numbers: No. NSC 102-2221-E-366-002, and NSC102-2221-E-110-029) and from the National Sun Yat-sen University (The Aim for the Top University Project, NSYSU)

REFERENCES

1. Jung JP, Lee JB, Kim JS, Park JS: Fabrication and characterization of high frequency SAW device with IDT/ZnO/AlN/Si configuration: role of AlN buffer. *Thin Solid Films*. 2004, 447–448:605-9.

2. Legrani O, Elmazria O, Zhgoon S, Pigeat P, Bartasyte A: Packageless AlN/ZnO/Si structure for SAW devices applications. *IEEE Sens J*. 2013, 13:487-91.

3. Meng X, Yang C, Chen Q, Gao Y, Yang J: Preparation of highly c-axis oriented AlN films on Si substrate with ZnO buffer layer by the DC magnetron sputtering. *Mater Lett*. 2013, 90:49-52.

4. Kao KS, Cheng CC, Chung CJ, Chen YC: Surface acoustic wave properties of proton-exchanged $LiNbO_3$ waveguides with SiO_2 film. *IEEE Trans Ultrason Ferroelectr Freq Control*. 2005, 52:503-6.

5. Wei CL, Chen YC, Cheng CC, Kao KS, Cheng DL, Cheng PS: Highly sensitive ultraviolet detector using a ZnO/Si layered SAW oscillator. *Thin Solid Films*. 2010, 518:3059-62.

6. Clement M, Olivares J, Iborra E, González-Castilla S, Rimmer N, Rastogi A: AlN films sputtered on iridium electrodes for bulk acoustic wave resonators. *Thin Solid Films*. 2009, 517:4673-8.

7. Lee JB, Jung JP, Lee MH, Park JS: Effects of bottom electrodes on the orientation of AlN films and the frequency responses of resonators in AlN-based FBARs. *Thin Solid Films*. 2004, 447–448:610-4.

8. Yim M, Kim DH, Chai D, Yoon G: Effects of thermal annealing of W/SiO_2 multilayer Bragg reflectors on resonance characteristics of film bulk acoustic resonator devices with cobalt electrodes. *J Vac Sci Technol A*. 2004, 22:465-71.

9. Kirby PB, Potter MDG, Williams CP, Lim MY: Thin film piezoelectric property considerations for surface acoustic wave and thin film bulk acoustic resonators. *J Eur Ceram Soc*. 2003, 23:2689-92.

10. Huang CL, Tay KW, Wu L: Fabrication and performance analysis of film bulk acoustic wave resonators. *Mater Lett.* 2005, 59:1012-6.

11. Umar A, Rahman MM, Vaseem M, Hahn YB: Ultra-sensitive cholesterol biosensor based on low-temperature grown ZnO nanoparticles. *Electrochem Commun.* 2009, 11:118-21.

12. Hong S, Yeo J, Manorotkul W, Kim G, Kwon J, An K, *et al.*: Low-temperature rapid fabrication of ZnO nanowire UV sensor array by laser-induced local hydrothermal growth. *J Nanomater.* 2013, 2013:246328.

13. Akiyama M, Morofuji Y, Kamohara T, Nishikubo K, Tsubai M, Fukuda O, *et al.*: Flexible piezoelectric pressure sensors using oriented aluminum nitride thin films prepared on polyethylene terephthalate films. *J Appl Phys.* 2006, 100:114318.

14. Iborra E, Clement M, Capilla J, Olivares J, Felmetsger V: Low-thickness high-quality aluminum nitride films for super high frequency solidly mounted resonators. *J Appl Phys* 2012, 520:3060-3.

15. Lin RC, Chen YC, Chang WT, Cheng CC, Kao KS: Highly sensitive mass sensor using film bulk acoustic resonator. *Sensor Actuat A-Phys.* 2008, 147:425-9.

16. Zhang H, Kim ES: Micromachined acoustic resonant mass sensor. *J Microelectromech S.* 2005, 14:699-706.

17. Chung CJ, Chen YC, Cheng CC, Kao KS: Synthesis and bulk acoustic wave properties on the dual mode frequency shift of solidly mounted resonators. *IEEE Trans Ultrason Ferroelectr Freq Control.* 2008, 55:857-64.

18. Schwindt CD, Settipane R: Allergic rhinitis (AR) is now estimated to affect some 1.4 billion people globally and continues to be on the rise. Am J Rhinol. *Allergy.* 2012, 26:S1-S1(1).

19. Rai M, Carpinella MC. Naturally occurring bioactive compounds. Baker & Taylor Books: Elsevier Science; 2006. pp. 271.

20. Chen YC, Chang WT, Cheng CC, Shen JY, Kao KS: Development of human IgE biosensor using Sezawa-mode SAW devices. *Curr Appl Phys.* 2014, 14:608-13.

21. Huang IY, Lee MC: Development of a FPW allergy biosensor for human IgE detection by MEMS and cystamine-based SAM technologies. *Sensor Actuat B-Chem.* 2008, 132:340-8.

22. Huang IY, Lee MC, Hsu CH, Wang CC: Development of a flexural plate-wave (FPW) immunoglobulin-E (IgE) allergy bio-sensing microsystem. *Sensor Actuat B-Chem.* 2012, 162:184-93.

23. Huang IY, Lee MC, Chang YW: *Development of a novel flexural plate wave biosensor for immunoglobulin-E detection by using SAM and MEMS technologies, 5th IEEE Conference on Sensors, October 2006.* IEEE, Daegu; 2006.

24. Huang IY, Lee MC, Chang YW, Huang RS: *Development and characterization of FPW based allergy biosensor, ISIE 2007. IEEE International Symposium on Industrial Electronics, June 2007.* IEEE, Daegu; 2007.

25. Qin L, Chen Q, Cheng H, Wang QM: Analytical study of dual-mode thin film bulk acoustic resonators (FBARs) based on ZnO and AlN films with tilted c-axis orientation. *IEEE T Ultrason Ferr.* 2010, 57:840-1853.

26. Chen YC, Chang WT, Kao KS, Yang CH, Cheng CC: The liquid sensor using thin film bulk acoustic resonator with c-axis tilted AlN films. *J Nanomater.* 2013, 2013:245095.

27. McNeil L, Grimsditch M, French RH: Vibrational spectroscopy of aluminum nitride. *J Am Ceram Soc.* 1993, 76:1132-6.

28. Ostuni E, Chapman RG, Holmlin RE, Takayama S, Whitesides GM: A survey of structure–property relationships of surfaces that resist the adsorption of protein. *Langmuir.* 2001, 17:5605-20.

29. Hook F, Rodahl M, Kasemo B, Brzezinski P: Structural changes in hemoglobin during adsorption to solid surfaces: effects of pH, ionic strength, and ligand binding. *Proc Natl Acad Sci U S A.* 1998, 95:12271-6.

30. Kaufman ED, Belyea J, Johnson MC, Nicholson ZM, Ricks JL, Shah PK, *et al.*: Probing protein adsorption onto mercaptoundecanoic acid stabilized gold nanoparticles and surfaces by quartz crystal microbalance and zeta-potential measurements. *Langmuir.* 2007, 23:6053-62.

31. Tang Q, Xu CH, Shi SQ, Zhou LM: Formation and characterization of protein patterns on the surfaces with different properties. *Synth Met.* 2004, 147:247-52.

32. Wang Y, Zhang Z, Jain V, Yi J, Mueller S, Sokolov J, *et al.*: Potentiometric sensors based on surface molecular imprinting:

detection of cancer biomarkers and viruses. *Sensor Actuat B-Chem.* 2010, 146:381-7.

33. Chung CJ, Chen YC, Cheng CC, Wang CM, Kao KS: *Superior dual mode resonances for 1/4 λ solidly mounted resonators, 2008 IEEE International Frequency Control Symposium, May 2008.* IEEE, Honolulu; 2008.

34. Beghi MG: Acoustic waves-from microdevices to helioseismology. *INTECH.* 2011, 14:501.

35. Link M, Schreiter M, Weber J, Primig R, Pitzer D, Gabl R: Solidly mounted ZnO shear mode film bulk acoustic resonators for sensing applications in liquids. *IEEE Trans Ultrason Ferroelectr Freq Control.* 2006, 53:492-6.

36. Yan Z, Zhou XY, Pang GKH, Zhang T, Liu WL, Cheng JG, *et al.*: ZnO-based film bulk acoustic resonator for high sensitivity biosensor applications. *Appl Phys Lett* 2007, 90:143503-1-3.

37. Weber J, Albers WM, Jussipekka T, Mathias L, Reinhard G, Wersing W, *et al.*: Shear mode FBARs as highly sensitive liquid biosensors. *Sensor Actuat A-Phys.* 2006, 128:84-8.

38. Wingqvist G, Bjurström J, Liljeholm L, Yantchev V, Katardjiev I: Shear mode AlN thin film electro-acoustic resonant sensor operation in viscous media. *Sensor Actuat B-Chem.* 2007, 123:466-73.

39. Mathias L, Weber J, Schreiter M, Wersing W, Elmazria O, Alnot P: Sensing characteristics of high-frequency shear mode resonators in glycerol solutions. *Sensor Actuat B-Chem.* 2007, 121:372-8.

40. Wingqvist G, Anderson H, Lennartsson C, Weissbach T, Yantchev V, Spetz AL: On the applicability of high frequency acoustic shear mode biosensing in view of thickness limitations set by the film resonance. *Biosens Bioelectron.* 2009, 24:3387-90.

41. Fua YQ, Luo JK, Du XY, Flewitt AJ, Li Y, Markx GH, *et al.*: Recent developments on ZnO films for acoustic wave based bio-sensing and microfluidic applications: a review. *Sensor Actuat B-Chem.* 2010, 143:606-9.

Angular Shaping of Fluorescence from Synthetic Opal-based Photonic Crystal

Vitalii Boiko[1], Galyna Dovbeshko[1], Leonid Dolgov[2], Valter Kiisk[2], Ilmo Sildos[2], Ardi Loot[2], and Vladimir Gorelik[3]

[1]Department of Physics of Biological System, Institute of Physics, NAS of Ukraine, Prospect Nauki 46, Kyiv 03680, Ukraine

[2]Laboratory of Laser Spectroscopy, Institute of Physics, University of Tartu, Ravila 14c, Tartu 50411, Estonia

[3]Raman Scattering Laboratory, P.N. Lebedev Physical Institute of the Russian Acad. Sci., Leninsky Prospect 53, Moscow 119991, Russia

ABSTRACT

Spectral, angular, and temporal distributions of fluorescence as well as specular reflection were investigated for silica-based artificial opals. Periodic arrangement of nanosized silica globules in the opal causes a specific dip in the defect-relate d fluorescence spectra and a peak in the reflectance spectrum. The spectral position of the dip coincides with the photonic stop band. The latter is dependent on the size of silica globules and the angle of observation. The spectral shape and intensity of defect-related fluorescence can be controlled by variation of detection angle. Fluorescence intensity increases up to two times at the edges of the spectral dip. Partial photobleaching of fluorescence was observed. Photonic origin of the observed effects is discussed.

BACKGROUND

Modification and enhancement of the fluorescence in photonic structures is important for development of optical sensors [1] and improvement of the fluorescence efficiency [2] and light harvesting ability in solar cells [3]. Most papers (see, for example, reviews [1],[4]) deal with experimental and theoretical aspects of the fluorophores embedded in the multilayered films forming one-dimensional (1D) photonic crystals. The behavior of such structures is determined by the interference of light that leads to decreased reflectance and enhanced fluorescence at certain directions of observation. Fluorescence has increased intensity and higher degree of polarization at these angles, in comparison to the background emission. The increase of light intensity at certain directions may be explained by resonant coupling of fluorescence with the waveguiding leaky modes in the 1D structure that can result in shorter fluorescent lifetimes and higher radiative rates [5],[6]. In particular, resonantly enhanced directional fluorescence with decreased lifetime was detected experimentally for a dye doped in 1D photonic crystals [5] and explained in terms of increased density of states near the photonic bandgap. Directional emission of light has also been reported for the multilayered films doped with quantum dots [7] and rare-earth ions [8].

Considerably less attention has been paid to the modification of light emission properties of fluorophores incorporated within three-dimensional (3D) photonic crystals, such as artificial opals composed of closely packed dielectric globules. Photonic stop zones in 3D structures have been already proved useful for blocking undesirable light emission [9]. In particular, this effect was used to avoid leakage of light from the dye-sensitized solar cells [3], to suppress the radiative channels[10], and in such a way to improve the Förster resonance energy transfer between the dye molecules situated inside the photonic structure [11].

In this work, we investigate the influence of photonic stop zones on the intrinsic fluorescence of 3D photonic crystal made of closely packed silica globules. Special attention is paid to the changes in the spectral shape of fluorescence as a function of the detection angle and the angular shift of photonic stop zone. It is demonstrated that the self-fluorescence of silica material can be enhanced at certain detection angles, near the spectral edges of the photonic stop zone.

METHODS

Preparation of photonic crystals based on synthetic opals was carried out in several steps. First, the silica globules were prepared by the hydrolysis of alkyl orthosilicate. Second, sedimentation and close-packing of these globules from the solution was achieved by centrifugation. Finally, the precipitated samples were annealed to obtain solid samples with size of several cm. Two samples (marked as 1 and 2) of compacted silica globules of slightly different sizes were selected for the study.

The samples were investigated by using optical and spectroscopic methods. SEM images of opals were obtained with EPMA SEI JXA-8200 microscope (JEOL. Ltd., Akishima-shi, Tokyo, Japan). Reflectance spectra at the normal incidence were measured on JASCO V-570 spectrophotometer (JASCO International Co. Ltd., Tokyo, Japan), whereas the angular dependence of reflectance was acquired on a custom goniometric setup [12]. Fluorescence was excited with Nd:YAG laser emitting at 266 nm. Fluorescence spectra were detected by means of Andor SR303i spectrograph equipped with a CCD camera.

RESULTS AND DISCUSSION

The electron micrographs of the samples (Figure 1) indicate rather uniform close-packing of nanosized silica globules (on the scale of several hundred micrometers at least). The average sizes of silica globules (determined from scanning electron microscopy (SEM) images) were 276 and 230 nm for the samples 1 and 2, respectively. According to the Raman and IR spectra [13], the globules consist of amorphous silica.

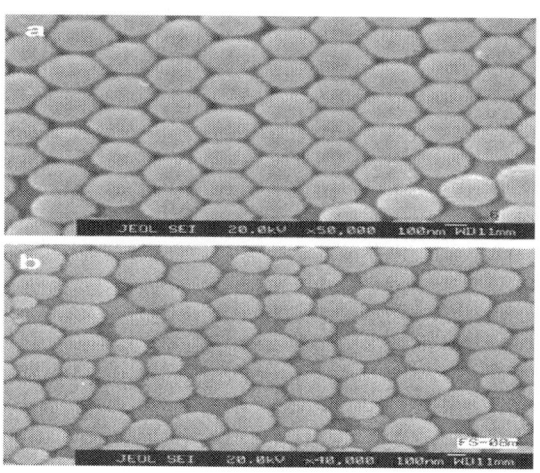

Figure 1: SEM image of silica globules. They are ordered in sample 1(a) and disordered in photonic glass (b).

Periodic spatial arrangement of silica globules results in distinct photonic properties that can be revealed in spectrally selective reflectivity. Indeed, the reflectance spectra of samples 1 and 2 (Figure 2) contain a single band of relatively strong reflectance with the peaks at 620 and 512 nm, correspondingly. The light cannot penetrate into the sample at these wavelengths, due to interference phenomena in such photonic structure.

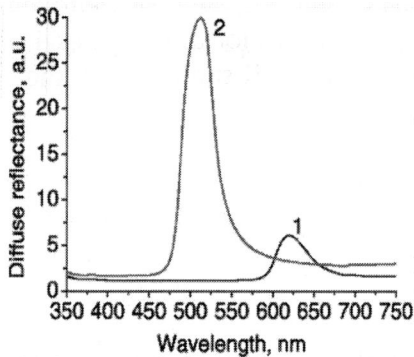

Figure 2: Spectra of light reflectance for photonic samples having different sizes of silica globules. Numbering of spectra corresponds to the numbering of samples.

The resonance condition depends on the direction of light propagation, and therefore, on the incidence/detection angle. The reflected light intensity was measured as a function of the detection angle for different wavelengths (Figures 3 and 4).

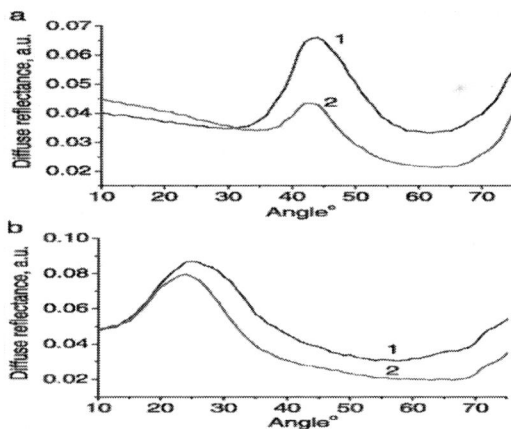

Figure 3: Angular dependence of light reflectance for sample 1.Wavelengths of incident light were 532 nm (a) and 593 nm (b). Both s-polarized (1) and p-polarized (2) light were tested.

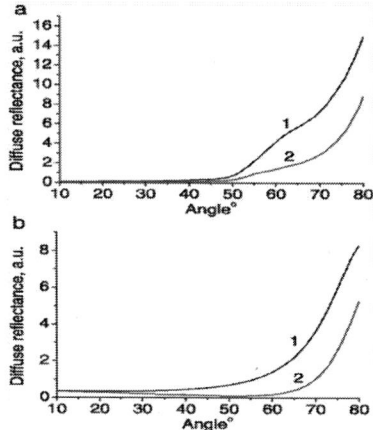

Figure 4: Angular dependence of light reflectance for sample 2. Wavelength of incident light were 402 nm (a) and 593 nm (b). Both s-polarized (1) and p-polarized (2) light were tested.

The incident light at 532 and 593 nm lies in the spectral range of photonic stop band for sample 1. Therefore, it is reflected stronger in the angular ranges from 35° to 55° (Figure 3a) and 15° to 35° (Figure 3b), respectively. In order to prove that the angular resonances depicted in Figure 3 will disappear outside the photonic stop band, we measured the angular dependence of light reflectance at 402 nm which is clearly outside the stop band range of the sample 1 (Figure 5). On the other hand, the wavelength 402 nm enters the stop band at the range of detection angles 50° to 70° in the case of sample 2 (Figure 4a).

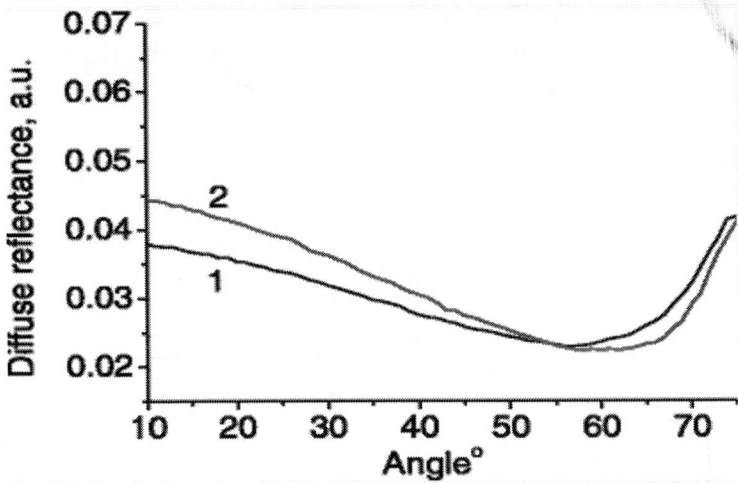

Figure 5: Angular dependence of light reflectance for sample 1. Wavelength of incident light was 402 nm. Both s-polarized (1) and p-polarized (2) light were tested.

In principle, one could employ the Brewster law for the estimation of effective refractive index of the samples, based on the angular dependences of reflectance, measured spectrally far from the photonic stop band (Figures 4b and 5). It was revealed, however, that the diffusely reflected light restricts this possibility.

Alternatively, the effective refractive index can be calculated using a modified Bragg law, which describes the relation between the local maximum in the reflectance spectrum λ_{max}, the corresponding angle θ of light detection, the size of globules D, and effective refractive index n_{eff} [14],[15]:

$$\lambda_{max} = 2a_{[111]} \sqrt{n_{eff}^2(\lambda) - \sin^2\theta}$$

(1)

where $a_{[111]} = \sqrt{\dfrac{2}{3}} D$ is the distance between the planes of closely packed silica globules in the direction [111].

For the sample 1, we have $\lambda_1 = 532$ nm, $\theta_1 = 46°$ (Figure 3a) and

$\lambda_2 = 593$ nm, $\theta_2 = 25°$ (Figure 3b). Substituting these values into the Equation 1 yields a system of two equations with two unknown parameters D and n_{eff}. The solution of this system gives $D = 276$ nm and $n_{eff} = 1.38$.

For the sample 2, we have $\lambda_1 = 512$ nm, $\theta_1 = 10°$ and $\lambda_2 = 500$ nm, $\theta_2 = 20°$ (Figure 6) yielding $D = 230$ nm and $n_{eff} = 1.38$.

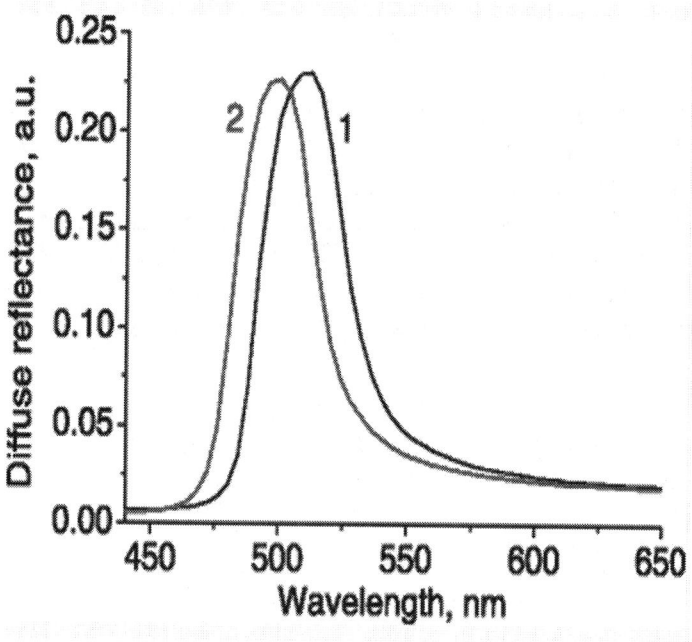

Figure 6: Light reflectance spectra for sample 2 measured at the angles 10° and 20°.

Obtained refractive indices correlate well with the literature data. Accurate description of optical properties of the mixed materials, particularly porous media, is not a trivial task. A list of possible approaches, by Maxwell-Garnett, Bruggeman and Lorentz-Lorenz models, Drude, or Silberstein formula, can be found in books, for example [16], and reviews [17],[18]. In our knowledge, none of the approaches mentioned above is general enough to account for the shape, size, and interconnection of pores in the real sample. Also, one can use for estimation of effective refractive index an equation known as "refractive mixing model" that is mathematically described by Birchak formula (Ref. [16], page 166). This model assumes that the refractive index of a composite mixture is an average of the indices of components weighted by their corresponding volumes:

$$n_{eff} = n_{globule}(\lambda) \cdot f + n_{inf} \cdot (1-f)$$

(2)

where $n_{globule}$ is a refractive index of silica, n_{inf} is a refractive index of the substance contained in the pores of photonic crystal ($n_{inf}=1$ for the air), and f is the volume fraction of globules in the sample ($f=0.74$ for dense hexagonal packing). Equation 2 was successfully applied earlier for the description of photonic crystals [14],[15], porous films [19], and soils [20]. Effective refractive index obtained for n_{eff} using Equation 2 was 1.39 [14] and 1.33 [15]. These values were obtained by assuming that $n_{globule}$ is equal to the refractive index of fused silica. In our opinion, it is a rough estimate, because silica nanoglobules are usually porous and, as a consequence, their refractive index could be slightly smaller than the refractive index of the fused silica.

Fluorescence spectra of samples excited with UV light ($\lambda_{exc}=266$ nm) consist of several broad, overlapping bands in the range of 400 to 700 nm (Figure 7).

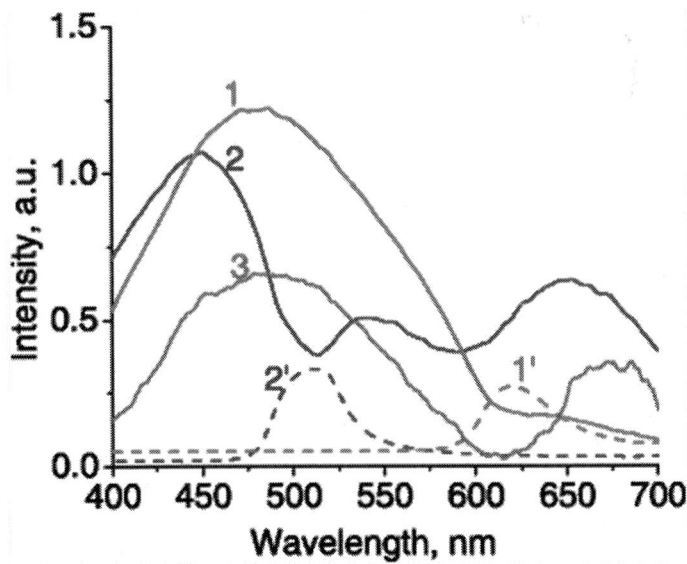

Figure 7: Fluorescence spectra of photonic crystal samples with different spectral positions of stop zones. Numbering of spectra 1 and 2 corresponds to the numbering of samples. Spectrum 3 is fluorescence of a soda lime glass slab. Spectra 1' and 2' are reflectance spectra of samples 1 and 2, respectively (plotted not to scale with fluorescence).

The spectra may belong to electronic transitions in the defect centers of -Si-O-Si-O- lattice which have the energies in the visible range [21]. Particularly, the maximum at 450 nm in spectrum 2 can be associated with twofold Si-oxygen deficiency center $O_2 = Si$:[22],[23]. Fluorescence bands with maxima near 500 nm in spectra 1 and 3 can be related to the hydrogenated $\equiv$ Si-H defects, which are formed by attaching H and OH groups to the disrupted $\equiv$ Si$\bullet$ and $\equiv$ Si-O$\bullet$ bonds [24]. Red luminescence in the range of 600 to 700 nm (spectra 2 and 3) can be associated with non-bridging oxygen hole centers [25] or OH groups on the surface of the silica [14].

The fluorescence emission from photonic crystals is subject to partial photobleaching. The brightness of fluorescence at the laser spot incident on the sample surface decreases essentially during the first minutes of irradiation and then changes more gradually (Figure 8). The blue fluorescence band with the maximum at 450 nm is bleaching faster than the red band with the maximum at 650 nm. As a consequence, fluorescence acquires a reddish tint at a longer exposure of the sample to UV light (Figure 8, inset).

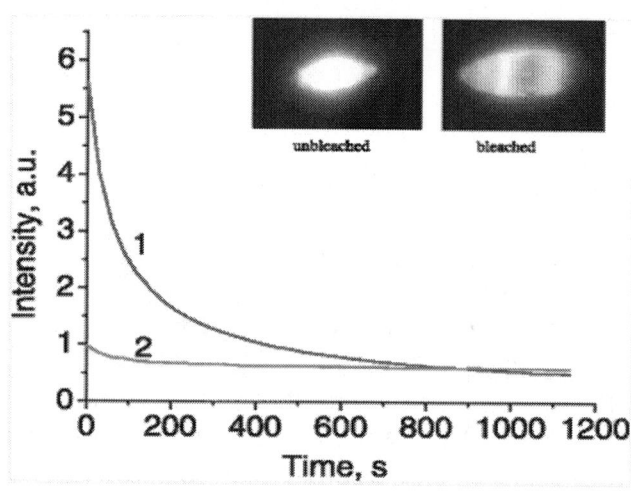

Figure 8: Photobleaching of photonic crystal fluorescence in time.Unequal kinetics of blue and red fluorescence is illustrated on the example of 450 nm (1) and 650 nm (2) spectral bands excited by steady laser irradiation at 266 nm. Images of the fluorescent spot on the surface of sample 2 in the initial and bleached states are shown in the inset.

Similar photobleaching of blue fluorescence has been reported earlier for the laser-treated silica waveguides [26]. This process may be caused by photoconversion of the silica twofold deficient centers into the E'-centers initiated by the two-photon absorption [27]. Further measurements were conducted after a preliminary UV irradiation of samples in the saturation region between 800 and 1,200 s (Figure 8), where the changes of fluorescence intensity caused by photobleaching are already minimal.

Fluorescence photobleaching may be affected by light confinement inside the photonic crystal and slow non-radiative migration of excitation between the defects in SiO_2 material. As a consequence, the excitation energy may decrease before reaching the defect-related fluorescence center. The longer the migration time, the smaller the energy reaching the luminescent center. Therefore, less energetic red fluorescence could become predominant after a prolonged irradiation. Similar effect and its origin have been discussed before [28].

An interesting feature in the recorded fluorescence is a dip in the spectrum 2 (Figure 7), with a minimum at 507 nm that is absent in spectra 1 and 3. This spectral feature overlaps with the maximum of reflectance, caused by photonic stop zone (Figure 7, spectrum 2'). Since the photonic stop zone of the sample 1 is almost outside the spectral range of fluorescence (Figure 7, spectrum 1'), the dip in fluorescence spectrum 1 is absent. The presence of the photonic dip in spectrum 2 demonstrates fundamental opportunity to control the spectral shape of fluorescence spectrum by 3D photonic structure of opal.

Recording fluorescence at different angles reveals a systematic shift of the abovementioned spectral dip. It shifts toward shorter wavelengths with the increase of detection angle (Figure 9). This allows to control the fluorescence intensity in the spectral range covered by the photonic stop zone by changing detection angle. The spectral position of photonic dip in the fluorescence can be described well with Equation 1.

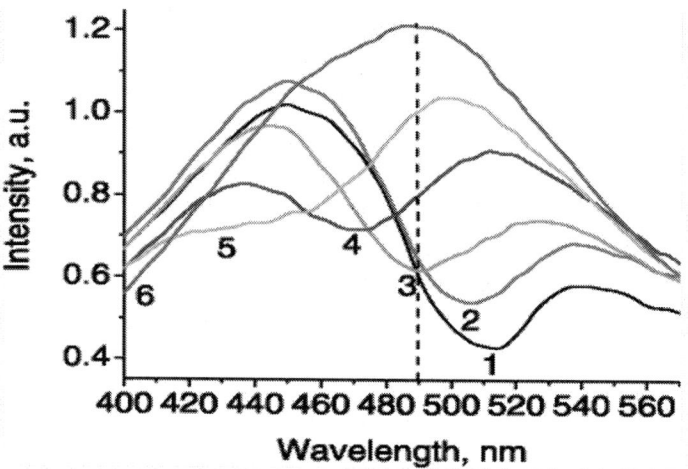

Figure 9: Fluorescence spectra of sample 2 measured at different detection angles. (1) 0°; (2) 10°; (3) 20°; (4) 30°; (5) 40°; (6) 70° ($\lambda_{exc} = 266$ nm).

In addition to the angular shift of the dip in fluorescence spectra, which is clearly of photonic origin, an enhancement of the fluorescence intensity was observed near the edges of spectral dip. For example, the fluorescence intensity at the detection angle of 70° is almost two times higher than that at the angle of 0° at the wavelength of 490 nm (marked by dashed line in Figure 9). We suppose that such angular enhancement in fluorescence also has photonic nature and is caused by blue shift of the photonic stop band at large angles. Similar effect was described for the light transmitted through a thin photonic crystal film [29]. The overlap of diffractional resonances associated with different systems of crystallographic planes would also lead to a redistribution of light intensity on the edges of the stop bands, which is visible for the transmitted light in Figures 2 and 3 in Ref. [29].

It should be noted that fluorescence intensity decreases at increasing of the observation angle for the fluorescent samples without a photonic superlattice, such as photonic glasses with disordered silica globules (Figure 10a) and reference slabs of soda lime glass (Figure 10b). Fluorescence emission experiences stronger refraction at high observation angles and higher reflectance at the air-glass interfaces, thus being unable to leave the sample (Figure 10a, inset).

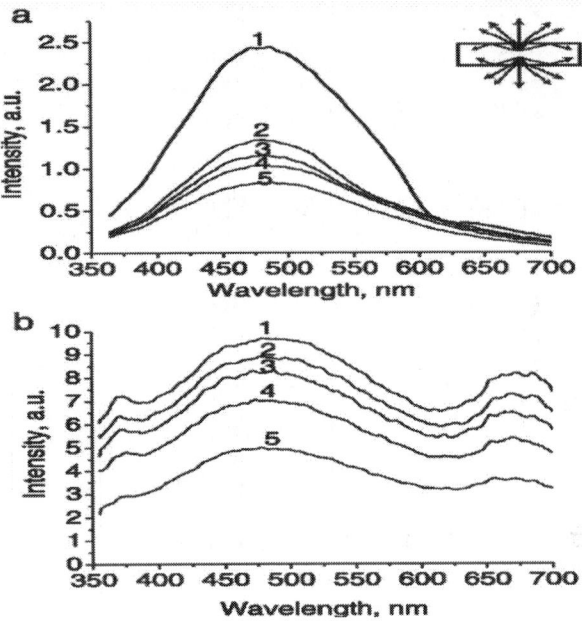

Figure 10: Fluorescence spectra of photonic glass (a) and soda lime glass slab (b), measured at different angles. (1) 0°; (2) 40°; (3) 50°; (4) 60°; (5) 70°. The inset depicts total internal reflection of light at high angles.

CONCLUSIONS

We demonstrated that intrinsic fluorescence of opal-based photonic crystals can be influenced by photonic stop zone. A decrease of the fluorescence at the wavelengths within the photonic stop band and its increase near the edges of stop band were observed. This effect could be proved by comparison of the fluorescent spectra detected at different angles, despite the undesirable photobleaching of the samples.

AUTHORS' CONTRIBUTIONS

VB and LD developed the idea of the work and were responsible for its realization. VG carried out the preparation and the necessary microscopic measurements of samples. VB, VK, and LD made the angular measurements of fluorescence. AL realized measurements of light reflection from the samples. GD, IS, and VG participated in the joint discussions with co-authors, explaining the scientific results. All authors have read and approved the final manuscript.

ACKNOWLEDGEMENTS

This work was supported by the Marie Curie ILSES project no. 612620, Russian-Ukrainian project 27-02-14, NATO SPS project NUKR. SFPP984702, Nanotwinning FP7 project (ID 294952), and European Regional Development Fund project TK114 and RFBR project 14-02-90406.

REFERENCES

1. Chaudhery V, George S, Lu M, Pokhriya A, Cunningham B: Nanostructured surfaces and detection instrumentation for photonic crystal enhanced fluorescence. *Sensors.* 2013, 13:5561-84.

2. Ganesh N, Zhang W, Mathias P, Chow E, Soares J, Malyarchuk V, et al.: Enhanced fluorescence emission from quantum dots on a photonic crystal surface. *Nat Nanotech.* 2007, 2:515-20.

3. Knabe S, Soleimani N, Markvart T, Bauer G: Efficient light trapping in a fluorescence solar collector by 3D photonic crystal. *Phys Status Solidi RRL* 2010, 4(5–6):118-20.

4. Barnes W: Fluorescence near interfaces: the role of photonic mode density. *J Mod Optic* 1998, 45(4):661-99.

5. Badugu R, Nowaczyk K, Descrovi E, Lakowicz J: Radiative decay engineering 6: fluorescence on one-dimensional photonic crystals. *Anal Biochem.* 2013, 442:83-96.

6. Calander N: Surface plasmon-coupled emission and Fabry-Perot resonance in the sample layer: a theoretical approach. *J Phys Chem B.* 2005, 109:13957-63.

7. Sainidou R, Renger J, Teperik T, Gonzalez M, Quidant R, Javier Garcia De Abajo F:Extraordinary all-dielectric light enhancement over large volumes. *Nano Lett* 2010, 10:4450-5.

8. Dolgov L, Kiisk V, Matt R, Pikker S, Sildos I: Tailoring of the

spectral-directional characteristics of rare-earth fluorescence by metal-dielectric planar structures. *Appl Phys B*. 2012, 107:749-53.

9. Baert K, Kolaric B, Libaers W, Vallee R, Di Vece M, Lievens P, *et al.*: Angular dependence of fluorescence emission from quantum dots inside a photonic crystal. *Res Lett Nanotechnol* 2008, 2008:4. Article ID 974072

10. Krishnamoorthy H, Song J, Kretzschmar I, Menon V. Photoluminescence modification in self-assembled fluorescent 3D photonic crystals, IEEE Applications and Technology Conference (LISAT). 2010. Conference Publication:1–4.

11. Kolaric B, Baert K, Van der Auweraer M, Valee R, Clays K: Controlling the fluorescence resonant energy transfer by photonic crystal band gap engineering. *Chem Mater.* 2007, 19:5547-555.

12. Loot A, Dolgov L, Pikker S, Lõhmus R: Goniometric setup for plasmonic measurements and characterization of optical coatings. In *Springer proceedings in physics 146: nanomaterials imaging techniques, surface studies, and applications*. Edited by Fesenko O, Yatsenko L, Brodin M. Springer, New York; 2013:119-34.

13. Dovbeshko G, Fesenko O, Boyko V, Romanyuk V, Moiseyenko V, Gorelik V, *et al.*: Vibrational spectra of opal-based photonic crystals. *IOP Conf Ser Mater Sci Eng* 2012, 38(1):12008-13.

14. Gruzintsev A, Emelchenko G, Masalov V, Romanelli M, Barthou C, Benalloul P, *et al.*:Luminescent properties of synthetic opal. *Inorg Mater* 2008, 44(2):159-64.

15. Reynolds A, Lopez-Tejeira F, Cassagne D, Garcia-Vidal F, Jouanin C, Sanchez-Dehesa J:Spectral properties of opal-based photonic crystals having a SiO$_2$ matrix. *Phys Rev B* 1999, 60(16):11422-6.

16. Sihvola A: *Electromagnetic mixing formulas and applications. IEE Electromagnetic Waves Series 47.* The Institution of Electrical Engineers, London; 1999.

17. Braun M, Pilon L: Effective optical properties of non-absorbing nanoporous thin films. *Thin Solid Films.* 2006, 496:505-14.

18. Hutchinson N, Coquil T, Navid A, Pilon L: Effective optical

properties of highly ordered mesoporous thin films. *Thin Solid Films*. 2010, 518:2141-6.

19. Taylor D, Fleig P, Hietala S: Technique for characterization of thin film porosity. *Thin Solid Films*. 1998, 332:257-61.

20. Mironov V, Dobson M, Kaupp V, Komarov S, Kleshchenko V: Generalized refractive mixing dielectric model for moist soils. *IEEE Trans Geosci Rem Sens* 2004, 42(4):773-85.

21. Salh R: Defect related luminescence in silicon dioxide network: a review. In *Crystalline Silicon: Properties and Uses*. Edited by Basu S. InTech, Rijeka; 2011:135-72.

22. Skuja L, Streletsky A, Pakovich A: A new intrinsic defect in amorphous SiO2: twofold coordinated silicon. *Solid State Commun* 1984, 50(12):1069-72.

23. Skuja L: Isoelectronic series of twofold coordinated Si, Ge, and Sn atoms in glassy SiO_2: a luminescence study. *J Non Cryst Solids*. 1992, 149:77-95.

24. Glinka Y, Lin S, Chen Y: The photoluminescence from hydrogen-related species in composites of SiO_2 nanoparticles. *Appl Phys Lett* 1999, 75(6):778-80.

25. Skuja L: Optical properties of defects in silica. In *Defects in SiO2 and related dielectrics: science and technology. NATO Science Series*. Edited by Pacchioni G, Skuja L, Griscom DL. Springer, Netherlands; 2000:73-116.

26. Albert J, Malo B, Johnson D, Hill K. Some optical properties of waveguides made by high energy ion implantation in fused silica, Conference Paper: Integrated Photonics Research New Orleans, Louisiana. 1992. ISBN: 1-55752-232-4

27. Imai H, Arai K, Imagawa H, Hosono H, Abe Y: Two types of oxygen-deficient centers in synthetic silica glass. *Phys Rev B* 1988, 38(17):2772-1275.

28. Goushcha A, Manzo A, Scott G, Christophorov L, Knox P, Barabash Y, *et al.*: Self-regulation phenomena applied to bacterial reaction centers. 2. Nonequilibrium adiabatic potential: dark and light conformations revisited. *Biophys J* 2003, 84(2):1146-60.

29. Romanov S: Anisotropy of light propagation in thin opal films. *Phys Solid State* 2007, 49(3):536-46.

Citations

CHAPTER 1

Songlin Duan, Pei Feng, Chengde Gao, Tao Xiao, Kun Yu, Cijun Shuai, and Shuping Peng, Microstructure Evolution and Mechanical Properties Improvement in Liquid-Phase-Sintered Hydroxyapatite by Laser Sintering, doi:10.3390/ma8031162.

CHAPTER 2

Gareth Edwards, Haijiang Li, and Bin Wang, BIM Based Collaborative and Interactive Design Process Using Computer Game Engine for General End-users, doi:10.1186/s40327-015-0018-2.

CHAPTER 3

Baboo Lesh Gowreesunker and Savvas A Tassou, Approaches for Modelling the Energy Flow in Food Chains, doi:10.1186/s13705-015-0035-y.

CHAPTER 4

Richard Zuech, Taghi M Khoshgoftaar, and Randall Wald, Intrusion Detection and Big Heterogeneous Data: A Survey, doi:10.1186/s40537-015-0013-4.

CHAPTER 5

Siripong Mahaphasukwat, Kazumasa Shimamoto, Shota Hayashida, Yu Sekiguchi, and Chiaki Sato, Mode I Critical Fracture Energy of Adhesively Bonded Joints between Glass Fibers Reinforced Thermoplastics, doi:10.1186/s40563-015-0036-2.

CHAPTER 6

Jamie Benson, Chung Man Fung, Jonathan Stephen Lloyd, Davide Deganello, Nathan Andrew Smith, and Kar Seng Teng, Direct Patterning of Gold Nanoparticles Using Flexographic Printing for Biosensing Applications, doi:10.1186/s11671-015-0835-1.

CHAPTER 7

Ying-Chung Chen, Wei-Che Shih, Wei-Tsai Chang, Chun-Hung Yang, Kuo-Sheng Kaoand Chien-Chuan Cheng, Biosensor for Human IgE Detection Using Shear-mode FBAR Devices, doi:10.1186/s11671-015-0736-3.

CHAPTER 8

Vitalii Boiko, Galyna Dovbeshko, Leonid Dolgov, Valter Kiisk, Ilmo Sildos, Ardi Loot, and Vladimir Gorelik, Angular Shaping of Fluorescence from Synthetic Opal-Based Photonic Crystal, doi:10.1186/s11671-015-0781-y.

Index

A

Adhesive bonding technology 166
Application Programming Interface (API) 42
Application Protocol-based Intrusion Detection System (APIDS) 99
Architectural, Engineering and Construction (AEC) 21
Attack Modeling and Security Evaluation Component (AMSEC) 128
Auger electron spectroscopy (AES) 193

B

Building Information Modelling (BIM) 21, 22, 23, 25, 48

C

Cloud Computing Storage System (CCSS) 116
Collaborative Intrusion Detection System (CIDS) 124
Common Information Model (CIM) 127, 161
Computer-aided design (CAD) 11

D

Data envelopment analysis (DEA) 76
Data Key Store (DKS) 117
Data Loss Prevention (DLP) 143
Deionised (DI) 190
Denial of Service (DOS) 145
Distributed Denial of Service

(DDOS) 145
Distributed Intrusion Detection System (DIDS) 125
Double cantilever beam (DCB) 165, 167
Dulbecco's Modified Eagle's Medium (DMEM) 14

E

Energy dispersive spectroscopy (EDS) 5
Energy dispersive X-ray (EDX) 193, 199
Energy efficiency indicator (EPI) 65
Enterprise Security Intelligence (ESI) 130

F

Fiber reinforced plastics (FRP) 166
Film bulk acoustic resonators (FBARs) 207
Full width at half maximum (FWHM) 214

G

Glass fiber reinforced thermoplastics (GFRTP) 167
Glucose oxidase (GOx) 188, 189
Gobal greenhouse gas (GHG) 57

H

Hadoop Distributed File System (HDFS) 116
Host-based Intrusion Detection System (HIDS) 99
Hyper Text Transfer Protocol (HTTP) 99

I

Index decomposition analysis (IDA) 80
input-output (IO) 68

International Energy Agency (IEA) 73, 95
Internet Protocol (IP) 42
Intrusion Detection Message Exchange Format (IDMEF) 124
Intrusion Detection Systems (IDSs) 101

L

Life cycle assessment (LCA) 60
Life cycle inventory (LCI) 60
Limit of detection (LoD) 190
Linear elastic fracture mechanic (LEFM) 167, 184
Liquid phase sintering (LPS 3
Log mean Divisia index (LMDI) 71
Low-pressure chemical vapor deposition (LPCVD) 210

N

Network Intrusion Detection System (NIDS) 99
Network Traffic Recording System (NTRS) 116

O

one-dimensional (1D) 228
Operation Technologies (OT) 139

P

Phosphate buffered saline (PBS) 190
Point-of-care (PoC) 189
Polyamide 6 (PA6) 165, 168
Poly-L-lactic acid (PLLA) 3
Polyvinylpyrrolidone (PVP) 190
Protocol-based Intrusion Detection System (PIDS) 99
Prototype system 22, 48
Psopropyl alcohol (IPA) 191
Publicly available specification (PAS) 61

R

Radio frequency (RF) 207, 210
Return On Investment (ROI) 139

S

Scanning auger microscopy (SAM)
193
Scanning electron microscope (SEM)
193
Scanning electron microscopy (SEM)
229
Security Information and Event Management (SIEM) 98, 160
Self-assembly monolayers (SAMs)
208, 212

Simulated body fluid (SBF) 2, 3, 8
Structural decomposition analysis
(SDA) 69, 71, 80
Structured Query Language (SQL)
99
Supply chain analysis (SCA) 69, 71
Surface acoustic wave (SAW) 208

T

Three-dimensional (3D) 228

U

Universal Plug and Play (UPnP) 50